Digital Innovation

Harnessing the Value of Open Data

Open Innovation: Bridging Theory and Practice

ISSN 2424-8231

The series aims to contribute to knowledge creation and more importantly, to knowledge accumulation, through the combination of multiple streams, perspectives, disciplinary approaches and diverse backgrounds. In doing so, it departs from the current body of literature adopting a purely academic perspective on Open Innovation, and thus restates the importance of anchoring Open Innovation research into the reality, practices, challenges facing firms and policymakers. This book series covers multiple perspectives, such as measuring and assessing the impact of Open Innovation, dealing with organizational matters and culture, designing strategies, policies, incentives and measures to support and implement Open Innovation, and discussing the advantages and limitations of adopting Open Innovation strategies.

Published

Vol. 4 *Digital Innovation: Harnessing the Value of Open Data*
edited by Anne-Laure Mention

Vol. 3 *Open Innovation and Knowledge Management in Small and Medium Enterprises*
edited by Susanne Durst, Serdal Temel and Helio Aisenberg Ferenhof

Vol. 2 *Open Innovation: Unveiling the Power of the Human Element*
edited by Anne-Laure Mention and Dimitrios G Salampasis

Vol. 1 *Open Innovation: A Multifaceted Perspective (In 2 Parts)*
edited by Anne-Laure Mention and Marko Torkkeli

Open Innovation: Bridging Theory and Practice Volume 4

Digital Innovation

Harnessing the Value of Open Data

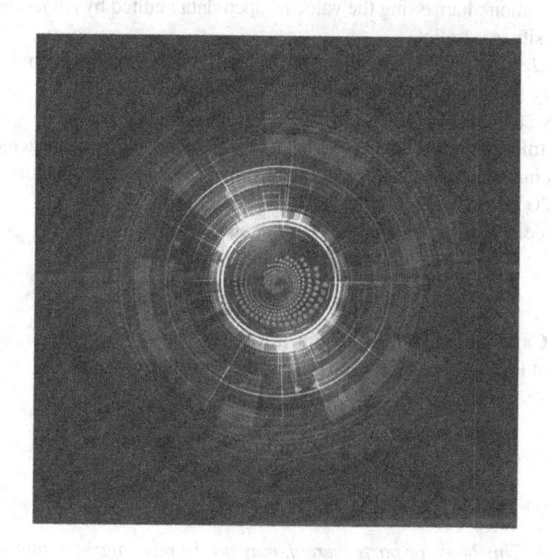

Editor

Anne-Laure Mention

RMIT University, Australia

World Scientific

NEW JERSEY • LONDON • SINGAPORE • BEIJING • SHANGHAI • HONG KONG • TAIPEI • CHENNAI • TOKYO

Published by

World Scientific Publishing Co. Pte. Ltd.
5 Toh Tuck Link, Singapore 596224
USA office: 27 Warren Street, Suite 401-402, Hackensack, NJ 07601
UK office: 57 Shelton Street, Covent Garden, London WC2H 9HE

Library of Congress Cataloging-in-Publication Data
Names: Mention, Anne-Laure, editor.
Title: Digital innovation : harnessing the value of open data / edited by Anne-Laure Mention (RMIT University, Australia).
Description: New Jersey : World Scientific, [2019] | Series : Open innovation: Bridging theory and practice ; Volume 4 | Includes index.
Identifiers: LCCN 2019008601 | ISBN 9789813271630 (hbk : alk. paper)
Subjects: LCSH: Information technology--Management. | Knowledge management--Technological innovations. | Open access publishing. | Transparency in government.
Classification: LCC HD30.2 .D535 2019 | DDC 658.4/038028557
LC record available at https://lccn.loc.gov/2019008601

British Library Cataloguing-in-Publication Data
A catalogue record for this book is available from the British Library.

For any available supplementary material, please visit
https://www.worldscientific.com/worldscibooks/10.1142/11026#t=suppl

Desk Editors: Anthony Alexander/Sylvia Koh

Typeset by Stallion Press
Email: enquiries@stallionpress.com

Printed in Singapore

About the Editor

Anne-Laure Mention is the Director of the Global Business Innovation Enabling Capability Platform at RMIT, Melbourne, Australia. She is also a Professor at the School of Management at RMIT, Melbourne; a Visiting Professor at Université de Liège, Belgium; the Deputy Head of the Centre d'Evaluation de la Performance des Entreprises; and a Visiting Professor at Tampere University of Technology, Finland. She holds several other visiting positions in Europe and Asia. Anne-Laure is one of the founding editors of the *Journal of Innovation Management*, and the Deputy Head of the ISPIM Advisory Board. She is the co-editor of a book series *Open Innovation: Bridging Theory and Practice*, published by World Scientific Publishing Co. Her research interests revolve around open and collaborative innovation, innovation in business to business services, with a particular focus on financial industry and fintech, technology management and business venturing. She has been awarded the prestigious IBM Faculty Award twice for her research on innovation.

About the Contributors

Hardik Bhimani is an Economic Strategist with over 10 years of expertise in corporate negotiations, digital change management and executive leadership. Hardik conducts multi-disciplinary and multi-method research at the intersection of psychology and management, with a focus on strategic cognition. His current projects explore cognitive processes in open innovation and socio-cognitive influences to decisions in sustainable supply chain management. Hardik is a regular contributor at International Society for Professional Innovation Management (ISPIM). He is also associated with the managerial cognition and open innovation interest groups at the Academy of Management (AoM) and behavioral strategy interest group at the Strategic Management Society (SMS). Hardik is pursuing a PhD in Management (Open Innovation) at RMIT University and is concurrently training for Organizational Psychologist registration in Australia. Hardik teaches strategy and economics courses and has recently accepted RMIT's Good Teaching award in recognition of his student engagement and practical theorizing.

Marcel Bogers is an Associate Professor of Innovation and Entrepreneurship at the University of Copenhagen. He works at the Unit for Innovation, Entrepreneurship and Management at the Department of Food and Resource Economics, Faculty of Science. He obtained an MSc from Eindhoven University of Technology and a PhD in Management of Technology from Ecole Polytechnique

Fédérale de Lausanne (Swiss Federal Institute of Technology). He previously worked (in permanent or visiting positions) at Chalmers University of Technology, University of Trento and University of Southern Denmark. His main interests center on the design, organization and management of technology, innovation and entrepreneurship. More specifically, his research explores openness and participation in innovation and entrepreneurial processes within, outside and between organizations. In this context, he has studied issues such as open innovation, business models, family businesses, users as innovators, collaborative, prototyping, improvisation and university–industry collaboration.

Terrence E. Brown is an Associate Professor of Entrepreneurship and Innovation at the Royal Institute of Technology, Sweden, where he is the Head of the Industrial Marketing and Entrepreneurship Division. He is the founding Editor of the *International Journal of Entrepreneurial Venturing*. Terrence holds a PhD in Entrepreneurship and Strategy from Rutgers University, which he completed in 1996. He also holds an MBA from Rutgers (obtained in 1994) as well as a second MBA from The Wharton School (obtained in 1986). He also holds an AB (Honors) from Brown University, which he obtained in 1984. Terrence is the former Dean of the Stockholm School of Entrepreneurship. He has also been a Visiting Professor at EBS Business School, Germany since 2005. He was also a Visiting Professor of Entrepreneurship and Strategy at Jönköping International Business School. His primary research area is value creation through the formation, management, rapid growth and rejuvenation of business enterprises. Terrence is a highly sought after speaker and has lectured throughout North America, Europe, Asia and Africa.

Sünje Dallmeier-Tiessen is the Data Coordinator in the Scientific Information Service at CERN. Together with her colleagues, she builds services to enable researchers to practice Open Science and to take steps towards reproducible research. Before she joined CERN, she worked for the Helmholtz Association in Germany.

She holds a PhD in Information Science from Humboldt University, Berlin, Germany.

Julius Francis Gomes is pursuing his PhD in International Business from the Oulu Business School, University of Oulu. His research focuses on futuristic business models for digital-intensive industries from the point of view of using business models as a mean to look into future of industries and understand business ecosystems. He is interested in researching business ecosystems in different contexts like cyber security, healthcare, telecommunications network, etc. with a business model perspective. He received his MSc in 2015 in International Business from the University of Oulu. Prior to that, he acquired an MBA in 2011, specializing in managing information systems in business applications. He has also enjoyed about three years in a top-tier bank in Bangladesh as a Channel Innovator.

John Gilchrist is a Senior Research Fellow at the Australian Catholic University School of Law. He studied Arts and Law at Monash University, and holds postgraduate degrees and qualifications from Monash University, the Queensland University of Technology, the Australian National University and the University of Canberra. He is a Fellow of the Higher Education Research and Development Society of Australasia. Dr Gilchrist has been involved with intellectual property as a policy and practicing lawyer and as an academic over a period of four decades. He began his professional life as a solicitor in the Australian Attorney General's Department and later with the Australian Government Solicitor's Office. His practical experience covers intellectual property (both in its international and national aspects), major government contracting and other commercial law fields. As a young lawyer, he was the Secretary of the *Copyright Law Committee on Reprographic Reproduction* (the Franki Committee) and more recently a member of the Copyright Law Review Committee on its *Crown Copyright* reference. He has published numerous articles on government copyright issues and is the author of the monograph *The Government and Copyright* published by Sydney University Press (2015), and the co-editor (and contributor), along with Professor Brian Fitzgerald, of the book *Copyright Perspectives*, published by Springer in the same year.

Marika Iivari is a Postdoctoral Researcher within Martti Ahtisaari Institute at AACSB-accredited Oulu Business School, Finland. She defended her doctoral dissertation on business models, open innovation and ecosystems. Her research has focused on inter-organizational and inter-industrial open innovation, public–private–people collaboration, and strategic management and governance of innovation ecosystems. Her special expertise is within future digital business and the digitalization of healthcare domain, the Internet of Things and ICT ecosystems, and smart cities as innovation platforms. Currently, she is serving as a Business Model Expert in the Digital Transition Partnership on the Urban Agenda for the EU.

Kati Lassila-Perini is a Project Leader at Helsinki Institute of Physics, Finland, and coordinates the Data Preservation and Open Access project in the CMS Experiment at CERN since its start in 2012. She is an experimental particle physicist with a PhD from ETHZ, Switzerland, with the subject of the discovery potential of the Higgs boson at the CMS Experiment. Kati is the prime mover in the open data release of CMS, and she has a particular interest in learning and teaching, and is happy to see the CMS open data in use in research, outreach and in education.

Sébastien Martin holds a Master's degrees in Urban Planning, History and Information Systems. Currently a PhD student in the field of Digital Humanities, he works on open data since 2012, with a particular interest on issues of barriers, ecosystems and value generation from open data. As an open data researcher at the Luxembourg Institute of Science and Technology, he is also involved in the BE-GOOD initiative.

Thomas McCauley started in particle astrophysics, getting his PhD from Northeastern University on the Pierre Auger experiment studying ultra-high-energy cosmic rays. He then moved to the IceCube experiment as a postdoctoral researcher at Lawrence Berkeley National Laboratory. In 2006, he joined CMS and is now a member of the University of Notre Dame group where he is also a member of QuarkNet, a research-based teacher professional development program. Tom works on data visualization, web applications, open data, public education and outreach and is looking for supersymmetry.

Anne-Laure Mention is the Director of the Global Business Innovation Enabling Capability Platform at RMIT, Melbourne, Australia. She is also a Professor at the School of Management at RMIT, Melbourne; a Visiting Professor at Université de Liège, Belgium; the Deputy Head of the Centre d'Evaluation de la Performance des Entreprises; and a Visiting Professor at Tampere University of Technology, Finland. She holds several other visiting positions in Europe and Asia. Anne-Laure is one of the founding editors of the *Journal of Innovation Management*, and the Deputy Head of the ISPIM Advisory Board. She is the co-editor of a book series Open *Innovation: Bridging Theory and Practice*, published by World Scientific Publishing Co. Her research interests revolve around open and collaborative innovation, innovation in business to business services, with a particular focus on financial industry and fintech, technology management and business venturing. She has been awarded the prestigious IBM Faculty Award twice for her research on innovation.

Bernardo Nicoletti is a Lecturer at the Università di TorVergata Rome, Italy and Temple University, Rome Campus. He serves as a Director in Transigma Emea, a strategy consultancy company specialized in process improvements and digitization in financial services with global assignments. Bernardo has been active in the financial services industry for several years. He applies an innovative approach of Lean and Digitize in his consultancy and has described the methodology in a book by Gower Press. His most recent books are *Mobile Banking, Financial Services and Cloud Computing, Digital Insurance and The Future of Fintech*, published by Palgrave-Macmillan. Bernardo worked with GE Capital as the Program Manager, as Group CTO of GE Capital, and with AIG as the CIO of Latin America.

Minna Pikkarainen is a joint Connected Health Professor of VTT Technical Research Centre of Finland and University of Oulu, Oulu Business School, Martti Ahtisaari Institute and Faculty of Medicine. As a Professor of Connected Health, Minna has been focusing on multi-disciplinary research on innovation management and ecosystemic business models in the context of connected health service cocreation. The research has been done together with VTT, Oulu University teams, companies, Oulu University Hospital and Central Finland University Hospital in the areas of preventive Mydata-based healthcare services, future pediatric surgery process and emergency care. During 2010–2012, Minna Pikkarainen was working as a Business Developer in Institute Mines Telecom, Paris and European Innovation Technology (EIT) network in Helsinki. Her key focus areas as a business developer was in healthcare organizations and digital cities. Previously, Minna's research was focused on the areas of software development, agile development and service innovation.

Jukka Ranta is a Researcher at the VTT Technical Research Centre of Finland. He has a licentiate of science degree in engineering from Helsinki University of Technology, with a major in systems analysis and operations research. His work has included industrial process simulation modeling and production planning, numerical optimal control, and safety of programmable automation systems. His current research interests include both qualitative and quantitative simulation modeling of socio-technical systems with emphasis on healthcare.

Achintya Rao is a science communicator working at CERN as a writer and web editor, currently also pursuing a PhD in science communication from UWE Bristol in the UK. At CERN, he has worked with CERN's IT and Scientific Information services as well as the CMS Collaboration's scientists to build and operate the CERN Open Data portal. He is particularly interested in the use of free and open-source software in the research process, and advocates for researchers moving away from proprietary tools as much as is possible.

Samuel Renault is a Researcher at the IT for Innovative Services (ITIS) department of the Luxembourg Institute of Science and Technology (LIST). He holds an engineering degree in Information Systems Architecture from the National Institute of Applied Science of Rouen, France. He is acting alternatively as an R&D Engineer, Data Scientist, Project Manager and Product Manager in the field of Procurement for various sectors including Logistics. In recent projects, he has developed and implemented a strategy for logistics-related data acquisition, transformation and representation. Samuel holds certifications and has experience in project management (HSPTP) and process assessment (TIPA® Lead assessor).

Jeffrey Roy is a Professor in the School of Public Administration at Dalhousie University's Faculty of Management. He is a widely published observer and critic of the impacts of digital technologies on government and democracy. He has worked with the United Nations, the OECD, multinational corporations and all levels of government in Canada. He has produced more than 80 peer-reviewed articles and chapters, and his most recent book was published in 2013 by Springer, *From Machinery to Mobility: Government and Democracy in a Participative Age*. Among other bodies, his research has been funded by the IBM Center for the Business of Government and the Social Sciences and Humanities Research Council of Canada. He may be reached at roy@dal.ca.

Tibor Šimko holds the PhD degree in Plasma Physics from the Comenius University Bratislava, Slovakia and from the University of Paris Sud, France. He works as a Computing Engineer at CERN where he founded the Invenio digital repository framework and worked as a Technology Director of INSPIRE, the high-energy physics information system. Tibor currently works as the Technical Lead of the CERN Analysis Preservation, the CERN Open Data and the REANA Reusable Analyses projects. Tibor's professional interests include open science and reproducible research, information management and retrieval, software architecture and development, psychology of programming, free software culture and more.

Serdar Temiz is a PhD candidate at the Industrial Marketing and Entrepreneurship Division of Industrial Economics and Management Department of KTH, Royal Institute of Technology and part of European Institute of Innovation and Technologies Doctoral program on ICT Innovation (eitictlabs.eu). His current research focuses on value creation with open data and he also does research on open innovation, open data, entrepreneurship, business models, virtual/mobile platforms, blockchain. He is the Chairman of Open Knowledge Sweden and also Open Knowledge Foundation's official representative as an Ambassador in Sweden. Open Knowledge Foundation is a worldwide non-profit network of people passionate about openness, using advocacy, technology and training to unlock information and enable people to work with it to create and share knowledge. Serdar has several years of experience as a technical assistant, software developer, telecom software engineer, system analyst, business developer, trainer and consultant. He is teaching/training/consulting topics that are related to entrepreneurship, innovation and technology at his own firm.

Slim Turki holds a Doctorate in Information Systems from the University of Geneva, Switzerland and in Computer Science from the Joseph Fourier University, Grenoble, France. As an open data researcher at the Luxembourg Institute of Science and Technology, his work is focused on data-driven services for smart cities. Since 2016, he is involved in BE-GOOD (Building an Ecosystem to Generate Opportunities in Open Data, Interreg NWE, http://www.nweurope.eu/begood) as the Lead Technical Partner. He has been involved in many national and international projects addressing multiple business sectors, such as financial services, e-government, tourism, mobility, education, employment and social services. He developed a solid experience in service analysis, design, re-engineering and compliance to regulations.

Peter Ylén is a Principal Scientist in the business, innovation and foresight research area at the VTT Technical Research Centre of Finland. He has worked at VTT since 2005 and before that as a Managing Director and Research Director in private sector as well as Researcher and Professor at Aalto University and University of Vaasa, respectively. His scientific interests are in system theory, service science, business ecosystems, systems thinking and system dynamics, modeling and simulation. He has worked in several healthcare sector EU projects, such as MIDAS (healthcare data and analytics), TBICare and Center TBI (traumatic brain injury) as well as nationally funded projects concentrating in topics including precision medicine ecosystem, cancer diagnostics, digital tools for preventive health and future hospital.

Contents

Introduction

Open Data as an Enterprising Revolution: Opinions and Insights

"There is no turning back the clock on our interconnected world, but we could jeopardize its benefits if we fail to invest in a trusted data environment".

— *Ellen Richey, Chief Enterprise Risk Officer, Visa*

Almost anything you do — from buying groceries, to buying your home, to managing your retirement investments and your health; most things your company or even your country does — is beset in a string of data — data about decisions, actions, assumptions and outcomes. It may not be an overstatement to say that data is the new oil! At best, data increases efficiency, motivates outcome-driven activities, helps identify what works, how and why; however, misused data can threaten our very existence. Used well, data can foster large-scale collaborative change, solve societal problems, expedite social inclusion, give power to citizens and encourage innovation, all in the quest for better human outcomes.

You need not look too far to find breathtaking claims of what data, especially the so-called "Big Data" and the analytics that decipher it can do. The world of information is changing, and to some, it already has! You are an integral part of it, whether you know it or not! Data is the new currency in the competitive world. Data is singular and plural. From industries to governments to NGOs, information management has evolved to gain competitive advantage, improve participation and bring about accountability. Data can be our savior but it is complex, dispersed, mostly unrelated and dynamic. Using data for social, environmental and economic impacts requires new age thinking. The challenges of intellectual property rights, risks of collecting, storing, distributing data, privacy and ethics remain as key concerns. Yet, whoever owns data can claim power over those who do not. One view is that data should be "open" to all without restrictions. Open data is a significant enabler of socio-economic development (e.g., education, health, welfare, productivity, etc.). Open data fosters open innovation.

However, *should data be open? How and when should it be open? Who is responsible for managing open data? What about intellectual property rights? Who is driving the "open data" movement?* These questions and many more on impact of opening up data from an organizational perspective, from social empowerment perspective and more generally from a human context are explored, investigated and explained in this evidence-based *Digital Innovation: Harnessing the Value of Open Data book*. It is written with the practitioner, manager, researcher and enthusiast in mind and is the next instalment of the series on *Open Innovation: Bridging, Theory and Practice.*

> *"There's nothing to be scared of in open data"*
>
> — Peter Corbett, CEO of iStrategyLabs

This book is for you if you ever wondered, *what is open data? Is the culture and competition shifting in a data-driven world? What are other agencies doing with open data? What are the way-points in opening up my firm's data? What challenges I might face and how others have tackled it? Are the governments supporting this? What can I harness from open data to support sustainable development?*

Open data is a revolutionary idea, where data should be freely accessible, carry no burden of copyright, patents and other modes of control in its use, and reuse. The premise being that openness fosters collaborative growth and innovation to tackle societal issues by leveraging otherwise closed and distributed knowledge. The logic of open data is not new, having been associated with the early years of European Enlightenment. The principles of open data have often drawn on the Merton thesis that science grows through improvements in methodologies, rooted in collective observations of a phenomenon in practice. Indeed, the first known peer-reviewed evidence-based scientific journal published in 1665, *Philosophical Transactions of the Royal Society*, required authors to embrace an open approach in scientific investigations, promoting replication, use and reuse of data as Boulton highlighted. Yet, the open data concept is relatively novel in innovation and management literature, having gained popularity with the advent of technological advancements, such as the World Wide Web, mobile-connectivity and more recently data-driven application programming interfaces (APIs).

Open data has been hailed in parallel to other "open" movements like open science, open innovation, open source, open government, open access and open content. Interestingly, the goals of open data are often in contrast with the intellectual property rights movements. On one hand, open data promotes fee and restriction-free access to textual (i.e., health reports, food safety reports, sustainability reports, etc.) and non-textual data (i.e., census, maps, weather, budget, geothermal data, etc.). The proponents of open data claim that any constraints and restrictions by law or strategy undermine creative collaboration and are against societal good. On the other hand, intellectual property advocates emphasize granting of legal monopolies by providing innovators with exclusive rights to exploit the innovation, excluding all others. They propose that licensing, copyright and patents provide an incentive for innovation, for exclusivity can lead to profitability, in turn encouraging innovators to invest in R&D activities. Thus, one school of thought fosters what the other prevents. This debate has found common places in board meetings where strategic choices of openness are on the agenda. Nevertheless, contemporary view of competitive strategy has often taken the middle ground. Modern strategists argue that salient among competition goals is the need to foster innovation and consumer welfare and that such an endeavor requires a shift from static economic policy based on constraints of supply and demand to a broader innovation perspective rooted in dynamic capabilities and absorptive capacity of the firm. These new age strategists acknowledge that not all data can be open, but some degree of openness is critical for the growth of the interconnected business–government–citizen sphere.

Despite this notion, to date most of the collaborative effort in opening up data can be attributed to government and public agencies, mostly because these agencies have superior access to generalized and factual data, provision of which is argued to least likely result in a "tragedy of the commons". Many positive arguments have been put forward by the likes of President Obama, the European Commission and other governments around the world. Common among these are that open data promotes efficiency through transparency and accountability. It is claimed that open data fosters self-empowerment, innovation and active public participation in decisions and actions that concern their welfare. To facilitate citizen integration into policy making, governments (e.g., UK, USA, Australia, etc.) have developed and implemented open data portals (i.e., data.gov.uk, data.gov, data.gov.au, etc.) to house federal, state and regional-level data. The concept is based on creative collaboration through knowledge sharing and the case is usually built around harnessing technological capabilities for practice through digital applications and services. However, impact is usually rooted in the transferability and connectivity of diverse and dispersed sets of data. The issue of interoperability of data remains a major obstacle for its utility. Under social and political pressure, agencies tend to release sets of unstructured data, which offer limited congruence to other datasets and hence lack utility for managerial sense-making. Indexing scattered data which is often in native language or laden with inconsistencies in terminologies and structure is the next big challenge in the quest for open data to elude the "tragedy of the anticommons".

It is generally held that public data belongs to the public and that facts cannot be copyrighted, yet cleaning, managing and disseminating data has a cost, which agencies would prefer to recover through fair remuneration. It is common in practice that despite the open data call, companies release a subset of datasets, target them to specific industry group suiting their strategic purpose or charge a fee for commercial use. Overall, open data phenomenon has captivated both sides of the society — those who are advocates of creative collaboration for societal progress and those who marshal privacy concerns and efficient use of resources for profit.

In 2011, the Open Government Partnership (OGP) was established which now has over 70 participating countries and 15 subnational governments with 2,500 commitments between them to make governments more transparent, accountable and open. In 2013, the Global Data Index (GDI) was created. The GDI examines how much open data is available in countries, allowing national communities to drive advocacy in their own contexts. It is important to emphasize that since its creation GDI has enabled collaboration and cross-communication between organizations and governments. Many countries have mandated public departments and agencies to release data without copyright restrictions to its distribution and reuse. In 2015, OGP launched a set of principles and best practices to facilitate the open data and open government movements, now housed within the International Open Data Charter (IODC). IODC advocates open data to empower citizens, promote the growth of small and medium-sized enterprises (SMEs) and create positive externalities through collaborative innovation efforts. However, as the chapters in this book will highlight, opening data is only a waypoint in the journey toward an integrated and encompassing open culture.

The first chapter "Open Data: Cliques to Culture" captures the basics of open data — the what, how and why. Logic of open data, the management perspectives and barriers to opening up are discussed with examples of current practices. It concludes with the view that open data is a purpose-driven phenomenon, laden with greater challenges than closed data, yet an opportunity toward a culture of transparency, accountability and enablement in a digital environment.

This journey has already begun. Yet, so far open data has taken more of a data provision perspective and less of a data integration perspective across the ecosystem. As Chapter 2 "Stimulation of Open Data Ecosystems: Learnings from Theory and Practice" highlights, the euphoric initial stage of open data movement has revealed little evidence of development and economic growth. The authors review public sector information experiences to presents a selection of success stories related to the development of local ecosystems. They amplify these insights to define the features of an environment conducive to open data as an instrument of capacity building and modernization.

Indeed, current scholarship lacks sufficient understanding of how diverse groups of dispersed stakeholders come together and actively participate in collaborative application of open data. In this view, Chapter 3 "The Ecosystem of Open Data Stakeholders in Sweden" links open innovation perspective to open data. The authors offer an insightful account of open government data, extending the stakeholder perspective to that of an organic ecosystem. Exploring the open data ecosystem in Sweden, they conclude that visualization, access, depositories and indexes are waypoints toward open innovation culture, one that requires further investigation into applications of various open data business models across stakeholder groups.

It goes without saying that we live in a digitized world. For businesses, digital transformation is both an advantage and a cultural challenge. Utility of open data business models as enabler of innovative services is rooted in firm's resilience for digital and technological transformations. Blockchain technology is the new age concept increasingly shaping current practices. Chapter 4, "Exploring the Digital Transformation via Open Data in Insurance", details a business model canvas and all its components. It stresses the importance of connected data and the need for governance. The author shares experiences of open data in retail and wholesale insurance services, concluding that despite its wide-ranging benefits, concerns for misuse, errors and equivocal production remain high, and further calls for capacity building to "exploit a treasure of data helping in getting a real competitive advantage".

On the governance and capacity-building perspective, Chapter 5 captures the concept of "fair dealing" in "Open Public Sector Information in Australia". This chapter written by the barrister and solicitor of the High Court of Australia and of the Supreme Court of the Australian Capital Territory provides an account of copyright in Australian government information and data. The author observes that salient changes in public sector information are not a result of copyright law, rather due to copyright practice. Of course, data that is sensitive or is confidential (e.g., material on national security and strategic implications) cannot be made open, yet author propels that logic of open data by suggesting that data that is made open in good faith should not be subject of legal action. More importantly, the chapter concludes with an emphasis on proactive support, training and education of officials tasked with releasing and managing open data.

Nevertheless, many governments are striving to develop open data strategies to empower their citizens, increase transparency and promote active participation in matters related to societal welfare. However, it is important to note that open data does not fit easily within the rubric of democratic governance and traditional public administration. Governments have been quite weary in terms of which data to put on display such that it demonstrates an openness to governance. In this view, Chapter 6, "From Open Data to Open Governance in Canada: Dissecting a Work in Progress" probes tensions between penchant for informational control and openness of information. The author highlights that within a broader democratic context, notions of individual privacy coexist uneasily with the emerging culture of openness and sharing. Roy concludes with a call for action — greater political innovation and dialogue to facilitate meaningful sharing of data in the quest for "a culture of responsible and genuine public involvement in the creation of public value". Effective political leadership is the essential prerequisite to openness. Furthermore, Roy calls for higher literacy of political leaders in digital matters by citing the Canadian context.

Addressing digital open innovation from a policy perspective, Chapter 7, "Toward Open Innovation and Data-Driven Health Policy Making" explores how open processes of knowledge flows can benefit healthcare policy makers and what could be done to foster open innovation in the healthcare sector. Decision-making in the healthcare sector has direct and indirect implications on individual patients, health professionals, health businesses as well as the society as a whole. Drawing attention to prevention rather than cure, authors provide empirically grounded findings of healthcare as an innovation system in Northern Finland. They find that open data-driven solutions, when coupled with supporting technologies, can provide faster and better use of data in decision-making, break up silos and speed up distribution of knowledge. Yet, this is possible only if the data is structured and has the characteristics to facilitate interoperability. As with previous chapters, authors emphasize the central role of policy makers as the leaders in advancing open innovation in the public sector. They conclude with a call for attention to policy maker's mindsets and tools that are able to "concretize this knowledge into actionable, usable relations and correlations".

In this view, the eighth and final chapter, "Early Experience with Open Data from CERN's Large Hadron Collider" provides practical insights on the motivation for opening up particle-physics data, challenges in doing so and the solutions developed to facilitate these efforts. For those who may not know, *CERN*, the European Laboratory for Particle Physics is the world's premier research facility for accelerator-based high-energy physics. The laboratory produces tens of petabytes of data to study the fundamental particles and forces of nature. To physicists and engineers, opening up this data has allowed for testing of theories and models to explain particle behavior, with applications to digital forensic studies in cloud computing, collaborative scientific investigations and knowledge preservation.

Overall, this book propels thought-provoking propositions in the case for open data, yet acknowledging the challenges and barriers in its practical applications. Awareness, engagement, training and education of those tasked in production, release and maintenance of open data remains an ongoing concern for researchers and practitioners. Besides, interoperability is the key to ensure effective utilization of data, for societal progress is rooted in how diverse and dispersed sets of data can be operationalized. Open data can improve efficiency and reduce costs, at least in the long run, with trust and trustworthiness important issues to be tackled through effective governance. Governments in this respect are central to the shift toward open culture, both for access to large volumes and diversity of data. Intrinsically, most scholars and practitioners agree that mindsets of leaders in politics and business are salient antecedents of how innovation takes shape and how open data is made available to others in the quest for improved services, products, policies, citizen-participation and self-empowerment.

We hope that you find the opinions and insights covered by some leading authors on open data to be pragmatic, challenging and practically useful in your journey toward an open culture — be it open science, open government or open innovation.

Chapter 1

Open Data: Cliques to Culture

Hardik Bhimani and Anne-Laure Mention

RMIT University, Australia

Data is arguably the new "oil". Cliques (or groups) of entities across the globe are shaping the organizational culture by steering operational transparency movements. Emphasis for societal progress has now shifted toward utilization of transferrable, trusted, standardized, equitable and freely accessible data. The premise being that such open data strengthens sustainability efforts through citizen empowerment, which in turn enhances conditions for civic engagement. Thus, a well-commanded open data-driven economy fosters openness, allowing translation of diverse sets of information into meaningful insights in governments, organizations, scientific centers and society. Open data may well be the central tenet of digital economy, innovative public services and effective governance, yet several behavioral, technical, economical and legal challenges remain. Opening up information resources, efforts to centralize user needs and monitoring of performance not only require sustainable funding & legislative enhancements, but also necessitates development of diverse skills to operate and prosper in the digital environment. This introductory chapter proposes open data as a purpose-driven phenomenon, laden with greater challenges than closed data, yet an opportunity toward a culture of transparency, accountability and enablement in a digital environment.

"The world is now awash in data and we can see consumers in clearer ways".

Max Levchin, PayPal Co-founder

"The goal is to turn data into information, and information into insight".

Carly Fiorina, ex-president and Chair of Hewlett-Packard

1. Introduction

These quotes capture much wisdom, in that they recognize the trend of data-driven decisions and, consequently the challenges it presents. Today firms increasingly associate data, a collection of facts, information and statistics, to competitive advantage through intertwining aspects of a "connected world". Managerial actions analogous with strategic decisions in production, pricing, supply chain, sustainable development, marketing, cost-management, service and many others are rooted in data (Gaudiano, 2017). Artificial intelligence (AI) and machine learning (ML) supported by information and communication technologies (ICT), have transformed the way we create, collate, communicate and cast-off information (Hartung *et al.*, 2010). In some respect, this phenomenon is not new (Boulton, 2014; Estermann, 2014). From agriculture to healthcare to commerce and justice, services that push and pull data to guide value chain activities have been fundamental to global development (Boulton, 2014). With technological movement, interactions between data and culture have centered on four data management capabilities — collection, collaboration, communication and construction — collectively guiding strategy-in-practice (Provost & Fawcett, 2013) for economic and social benefits through greater transparency and accountability (Davies & Perini, 2016; Pabón *et al.*, 2013; Zuiderwijk *et al.*, 2014).

Almost 20 years ago, Usama Fayyad, Gregory Piatetsky-Shapiro and Padhraic Smyth published "From Data Mining to Knowledge Discovery in Databases" which acknowledges extraction of data (i.e., data mining) as a historical tradition. Yet, blind application of algorithms has been known to guide value-destroying activities (Sayogo *et al.*, 2014), caused due to meaningless and confounded patterns (Fayyad, Piatetsky-Shapiro & Smyth, 1996). Over the last two decades, statisticians and computer scientists have progressively converged to convert data into information and information into insights and knowledge (Davenport & Harris, 2007; Hossain, Dwivedi & Rana, 2016). Data-driven insights forge better strategic decisions — this is now a well-acknowledged fact (Hossain, Dwivedi & Rana, 2016), mostly because it facilitates new knowledge (Shadbolt *et al.*, 2012) and allows managers to de-bias decisions in promoting evidence-based actions (McAfee, Brynjolfsson & Davenport, 2012; Rohunen *et al.*, 2014).

Recent public interests and developments in citizen-centered governance (Cerrillo-i-Martínez, 2012), rise in social media interactions with increasing dependence on mobile network (Huijboom & Van den Broek, 2011) and computer skills in relation to data creation and management (Boulton, 2014) have fueled the argument that data should be free and easy accessible by all (Hossain, Dwivedi & Rana, 2016). Data that is freely accessible online, available without restrictions and provided under open and unrestricted access to all is dubbed as "open data" (Huber, Rentocchini & Wainwright, 2016). Open data strategy entails organizations to release non-personal objective and factual data, collected in the process of its end-to-end operations, to anyone without restrictions (Bertot *et al.*, 2014; Kassen, 2013). The trend of open data is on the rise (Manyika *et al.*, 2013), in turn influencing company strategies and lives of people on a daily basis (European Commission, 2017a).

2. Open Data: A Conceptual Definition

Since the concept gathered steam on a global scale in 2009, with first the US and then the UK government embracing "open government" initiatives (Lathrop & Ruma, 2010; Meijer, Conradie & Choenni, 2014), several local, regional and national agencies have joined the cause of the "next big thing" in economic development — that is the open data philosophy. For our purposes, we define open data as:

> Open data is a set of facts, information or statistics that can be freely accessed, used, re-used, and redistributed by anyone; be reconfigurable to suit the organizational and environmental attributes, allowing interoperability through componentization and openness of otherwise intertwined complex systems.

This definition captures the central tenets of open data, being: (i) ease of availability and access, (ii) reuse, redistribution and reconfiguration and, (iii) interoperability. Ease of availability and access is critical to foster economic and scientific growth through knowledge-sharing activities (Davies & Perini, 2016). Reuse, redistribution, and reconfiguration of data suited to organizational and environmental context allows for initiatives forging new economic outputs through innovation (Carrara *et al.*, 2015; Guerin, 2014; Hossain, Dwivedi & Rana, 2016; Kundra, 2012; Petrov, Gurin & Manley, 2015). Indeed, arguments made in favor of economic benefits are the key drivers of open data initiatives in government-led and private innovations (Davies, Perini & Alonso, 2013; Mayinka *et al.*, 2013; Vickery, 2011). Interoperability in this regard refers to the suitability or "fit" of diverse datasets to meld, such that it enables componentizing and construction of innovative systems in the process of value-creation (Rothenberg, 2012). This interoperability is essential to create meaningful patterns for commercial applications in a cost-effective manner (McAfee, Brynjolfsson & Davenport, 2012).

2.1. *Open data: Value and benefits*

It is not surprising that the phenomenon of open data is transforming the way of life by promoting public and private value-creating activities (Manyika *et al.*, 2013; Obama, 2012; OECD, 2016). For instance, in 2016, there were over 30,000 downloads from Singapore's open data platform data.gov.sg, and about 2 million API calls per month from the developer's portal. As of January 2017 (Web Foundation, 2017), this one-stop portal provided access to more than 900 high-quality datasets from 70 public agencies and enabled sharing among different entities within Singapore (Phoensight, 2016). What started as cliques trying to find new productivity paths by segmenting markets centered on proprietary data is increasingly transforming the organizational and societal culture (Boulton *et al.*, 2011; Hartung *et al.*, 2010; Janssen, Charalabidis & Zuiderwijk, 2012; Linders, 2013; O-Hara, 2012). It is now an accepted argument that open data benefits customers and employees through improved insights, empowers institutions (public and private) through enhancements in transparency, and enhances the quality of life by stimulating innovations in products and services (Reynolds, 2017; Borzacchiello & Craglia, 2012; Kucera *et al.*, 2015).

2.2. Open data: Challenges and risks

The essence of open data implies moving from closed to open systems, affecting feedback loops where user input can drive strategy and policies (Jackson, 2003). The roles of data collector, user, distributor and manager are often blurred in open data management, creating complexity and challenges at project, institutional, local, national and technical levels (Zurada & Karwowski, 2011; Zuiderwijk, Janssen & Choenni, 2012; Zuiderwijk *et al.*, 2012). Open data management requires a fundamental shift from machine-like control to an evolutionary perspective, which guides development of behaviors as they emerge through interactions of otherwise independent network agents (Zuiderwijk, Janssen & Davis, 2014). Consequently, in embracing open data, managers (and legislators) find themselves amid networks that recognize the benefits of open data but at the expense of centralized control (Janssen, Charalabidis & Zuiderwijk, 2012).

In the remainder of this chapter, we introduce open data as a phenomenon that is driven by motives beyond supply and demand of data. We initiate a dialogue on the current challenges and risks in open data strategies and conclude with insights on business model innovations found in practice, therein proposing avenues for future developments in the discourse.

3. Open Data: As a Purpose-Driven Phenomenon

Open data is considered an enormous opportunity (Estermann, 2014) with advantages for a broad group of stakeholders (Zuiderwijk *et al.*, 2012). Research from the McKinsey Global Institute suggests that over $3 trillion in annual value could arise from the use of open data in applications across several domains of the global economy (Manyika *et al.*, 2013). In Europe, the size of open data economy is estimated between €27 billion and €140 billion (Kulk & Van Loenen, 2012). However, application of open data tends to accentuate the tendencies in decision-making to focus on purpose of openness within engaged domains of application to realize benefits (Davies & Perini, 2016). In other words, to explore the question of how open data influences practice, we need to understand the *why-how-what* frame. Figure 1 captures this logic of open data translation from conceptualization to impact. In this context, *why* refers to the drivers of open data, *how* refers to the decision frames or domains which facilitate the translation of open data into insights and *what* refers to the possible outputs in the application of open data strategy. Impact of open data is then reflected in the validity and assessment of realized innovative benefits in terms of economic, environmental, societal, health, cultural and academic outcomes.

The presence of inspiring exhibits and (experimental) insights tend to drive the behavior of observers in open data landscape. For instance, UK's *Show us the way* initiative was the driver for *MashUp Australia* (Huijboom & Van den Broek, 2011). In fact, political leadership from the top-down has been instrumental in raising awareness and urgency for application of open data policies. It was US President Obama's inaugural memorandum on open data that paved the path for UK's Prime Minister Gordon Brown to initiate the launch of

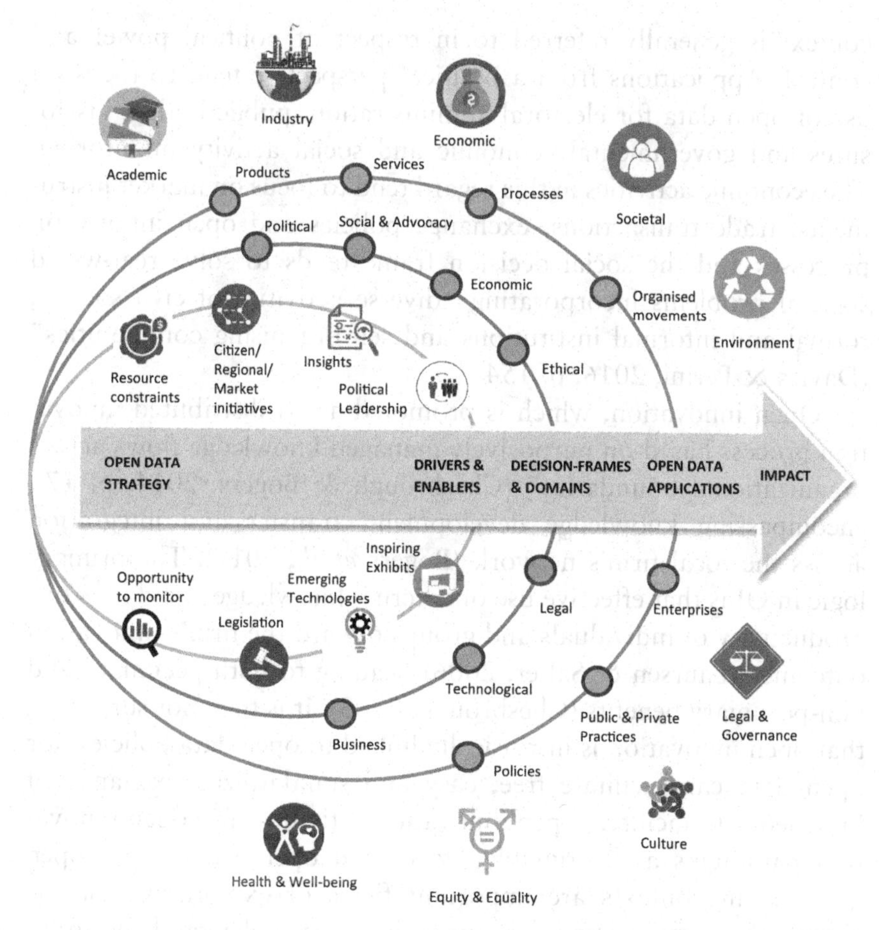

Fig. 1. Logic in open data philosophy from conceptualization to impact

data.gov.uk, a website that houses various local and national datasets for free use and reuse. Following such suit, the Australian government has created an open data toolkit and launched datastart, a public–private partnership that provides Australian tech start-ups with open innovation opportunities by facilitating the use of open government data (Australian Government, 2017). Political leaders of 75 countries now embrace open data policies and lead the change from top-down (Open Government Partnership, 2017). In the current literature, political decision frame or domain in open data context is generally referred to in respect of political power and control. Applications from a political perspective tend to focus on use of open data for electoral administration, public budget disclosures and governmental economic and social activity monitoring. The economic activities in this regard tend to focus on market instruments, trade transactions, exchange policies and open innovation processes and the social decision frame tends to solve real-world societal problems incorporating "diverse activities of civil society, formal and informal institutions and self-organising communities" (Davies & Perini, 2016, p. 154).

Open innovation, which is promoted as "a distributed innovation process based on purposively managed knowledge flows across organizational boundaries" (Chesbrough & Bogers, 2014, p. 17), encompassing knowledge development, transfer and integration across the focal firm's network (Bogers *et al.*, 2017). The primary logic in OI is that effective use of external knowledge could improve productivity of individuals and groups toward the firm's innovation outcomes (Laursen & Salter, 2006), leading to both pecuniary and non-pecuniary benefits (Chesbrough, 2003). It is then not surprising that open innovation is intrinsically linked to open data policies, for open data can facilitate free, easy and standardized exchange of knowledge to facilitate open innovation activities. The open innovation paradigms and, arguably, its related open science and open government contexts are important fields of exploration for the claims of potential impact of open data. To understand the influences of open data, it is important to explicitly consider the domains or decision frames that underline the decision-making process. For instance, the political and governance decision frames that have recently gained exposure through scientific inquiries include legislative and judicial processes, smart cities, access to scientific methodologies in delivery of public services, technological innovations and societal welfare. It is also important to note that each decision frame involves a different (and sometimes overlapping) constellation of stakeholders, concerns and disciplinary perspectives. Purpose of the open data policy or strategy is thus important to predict the trajectories of impact in research settings.

Purpose in this regard is intrinsically related to the source of data and its openness. While the purpose in some economies may be to maintain power by acting as the central node of information (Peled, 2011) with a focus on "open government" (Bichard & Knight, 2012; McGee & Edwards, 2016); other economies may rely on a broad community of organizations in creation and collaboration of data with a focus on "open society" (Fung & Weil, 2010). Open data thus may be of several types — (i) primary data (e.g., population statistics), (ii) secondary data (e.g., retail trends), (iii) real time (e.g., weather), (iv) location specific (e.g., geo-thermal activity), (v) commercial (e.g., healthcare budget, library data), (vi) welfare related (e.g., public housing reports, food safety reports), (vii) scientific (e.g., clinical results), (viii) generic (e.g., maps, satellite imaging, pictures) and so on (Rohunen *et al.*, 2014). Researchers have accordingly taken various perspectives to explain the purpose of open data (Hossain, Dwivedi & Rana, 2016).

4. Perspectives in Management of Open Data

Broadly, management practice in the context of open data can be categorized into four distinct but integrated perspectives — technical, social, political and organizational. The technical perspective in open data management is intrinsically linked to standardization and interoperability of data (Behkamal *et al.*, 2014; Borglund *et al.*, 2014; Alexopoulos, Spiliotopoulou & Charalabidis, 2013; Whitmore, 2014). From this viewpoint, technological innovations coupled with behavioral intentions are the key drivers of open data in policy and organizational strategy (Behkamal *et al.*, 2014; Lassinantti, Bergvall-Kåreborn & Ståhlbröst, 2014). The social perspective guiding behavioral intentions of decision-makers extends the utility of open data for the improvement of community services (Alexopoulos, Spiliotopoulou & Charalabidis, 2014; Bichard & Knight, 2012) through assimilation of civic information across various mediums of exchange (Garbett *et al.*, 2010). While some social initiatives recognize individual consumers as the means to an end (e.g., open data initiatives developed to commercialize and market public goods), some others recognize citizens as an end in themselves (e.g., initiatives that support collective action for betterment of social value through provision of open data on public goods and services). Accordingly, linear path of value creation in application of open data (Davies, 2010; Davies & Perini, 2016) is beset with economic challenges (Childs *et al.*, 2014). Drawing comparisons with open sourcing, economic perspective then refers to open data as a value-creation mechanism to improve profits (Lindman, 2014) through innovation (McLeod, 2012). Proponents of value creation consider open data as an enabler of people's power over their government (Tolbert & Mossberger, 2006), despite the inherent issues of privacy, security and trust on data (Bates, 2014; Meijer, Conradie & Choenni, 2014) that tend to make the process of releasing data to the public complicated (Zuiderwijk *et al.*, 2012; Zuiderwijk & Janssen, 2014).

Interestingly, government agencies are increasingly collaborating with universities and communities to develop and publicly release meaningful datasets for the purpose of securing financial budgets (Childs *et al.*, 2014), to attract new funds (Tananbaum, 2008) and to promote reuse of raw data (Hester, 2014). Organizational perspective in this regard refers to open data as an internal resource, which guides policy and strategy (Conradie & Choenni, 2014; Dulong de Rosnay & Janssen, 2014) toward institutional and societal value-creation activities (Krumholz & Waldstreicher, 2016). In this view, opening private data reduces information asymmetry in the market by allowing consumers to verify organization's sustainability and ethical practices (Eisenhardt, 1989), a strategy motivated to improve consumer trust (Carrigan & Attalla, 2001).

5. Motives in Open Data Policy and Strategy

In framing open data, it is important to acknowledge that strategic decision-making that guides the open data purpose perspective is a multi-level and multi-faceted process (Dunleavy *et al.*, 2006). It encompasses the quadruple helix model of government, research centers, organizations and citizens (Caragliu, Del Bo & Nijkamp,

2011; Maier-Rabler & Huber, 2012) working collaboratively toward innovative products and services. The context under which open data is applied (i.e., domain) is thus an important consideration in decision-making. In other words, how value is created from the use of open data guides the implications for its impact in practice (Gerunov, 2015; Srimarga *et al.*, 2014). Politically motivated purview tends to focus on open government with retention of power and control being the central tenets guiding open data projects (Davies & Perini, 2016). In fact, it is argued that it is politicians who institutionalized the open data concept (Huijboom & Van den Broek, 2011; Lathrop & Ruma, 2010).

Nations where political leaders view open data as a mechanism to engage citizens (Jansen, 2012) have far exceeded those where leaders do not share the same motivation toward open data. However, governments tend to align policies in response to social trends. Recent trends in social media-driven innovation (Roberts & Piller, 2016; Lin, Li & Wang, 2017) have motivated firms to embrace open data in development of services and products (Huijboom & Van Den Broek, 2011). Social motives in this regard tend to capture elements of civic imbalances (McGee *et al.*, 2015) and are often intertwined with economic value-creation through formal and informal networks engaged in exchange and trade (Krumholz & Waldstreicher, 2016; Lindman, 2014). Evidently, institutional pressures to release data to the public are associated with demands for reducing consumer information asymmetry (Castaldo *et al.*, 2009) to guide ethical consumption (Carrigan & Attalla, 2001; Sayogo *et al.*, 2014). Overall, the practices across various domains tend to demonstrate a distinction between following outcome-driven motivations in open data management.

5.1. *Increase participation, self-empowerment and engagement*

Common among open government projects (Zuiderwijk & Janssen, 2014), publishing of civic data empowers citizens to take control of decisions affecting their rights, for "a more informed citizen is a more empowered citizen" (Chief Secretary to the Treasury, 2009, p. 25). Since 2009, and subsequent to President Obama's commitment to government transparency and open government initiatives through launch of websites like www.recovery.gov and US data.gov, 56 countries had adopted open data philosophy by 2013 (Sayogo *et al.*, 2014). This number has now grown to 75 countries and 15 subnational governments who have collectively made 2500 commitments for open government and accountability (Open Government Partnership, 2017). The primary motivation being to engage public participation in democratic decision-making (Harrison & Sayogo, 2013) and engage a broad community of stakeholders in collective economic efforts (Linders & Wilson, 2011). Generally, social participation is a part of the broader strategy aimed at leveraging the advancements in information and communication technologies (ICT). For instance, engagement in Web 2.0 enabled social media (Bugshan, 2015) users to actively participate in the development of products and services (Kassen, 2013), including:

- Exercising rights to governmental budgets (Chief Secretary to the Treasury, 2009).
- Planning and use of common resources (i.e., water, air) and public goods (i.e., defense, public information).
- Product and service design and delivery (Ubaldi, 2013).
- Real-time collaboration in local, regional and nation-building activities.
- Policy making and dispute resolution (Ubaldi, 2013).
- Monitor government and private entity performance (Huijboom & Van Den Broek, 2011).

Open data coupled with timely and standardized formats can allow users to make informed decisions affecting public issues, such as healthcare (Krumholz, & Waldstreicher, 2016), education (Huijboom & Van den Broek, 2011), law enforcement (European Commission, 2003), to name a few. While much of the emphasis in practice has been on opening non-privacy-related and non-confidential data by government agencies (Kalampokis, Tambouris & Tarabanis, 2011;

Noveck, 2009), the increase in awareness has prompted private organizations to respond to stakeholder pressures for transparent, accountable and ethical practice (Baue & Murninghan, 2011; Doorey, 2011). The network effect of open data now extends to the creation of ecosystems (Hendler & Berners-Lee, 2010) nurturing a culture of collaborative data creation, accumulation and dissemination process, enhancing cost effectiveness and building capabilities at community and regional levels (Huijboom & Van den Broek, 2011; Ubaldi, 2013). The fundamental motivation in opening private data is that open data will improve efficiencies and effectiveness in consumer markets, provide opportunities to develop innovative tools and collate purposive market insights, thus enhancing customer value-creation efforts in the quest to achieve competitive advantage (Sayogo *et al.*, 2014; Sunstein, 2011). In fact, broadening the scope of organizational activities to include stakeholder interests traces back to origins of stakeholder theory (Baue & Murninghan, 2011). Yet, the momentum, level of engagement and tactics have changed in recent times in an effort to leverage technological capabilities in building collaborative partnerships (Baue & Murninghan, 2011).

5.2. *Promote product and service innovation*

Governments in their capacity as a "data provider" (O'Reilly, 2011) often seek to leverage the greater external (or private) capacity to innovate (Hippel & Krogh, 2003). The creation of open government to foster a culture of collaborative practice and stimulate industries has been the primary motivation cited in most government initiatives to date. For instance, UK's *Putting the Frontline First: Smarter Government* initiative states that open data can bring economic benefits to people and business by releasing "untapped enterprise and entrepreneurship" (Huijboom & Van den Broek, 2011, p. 4). Likewise, Denmark's *Open Data Innovation Strategy* initiative promotes the utility of open data as a creative practical solution to "create new business in developing digital services and advanced content" (Huijboom & Van den Broek, 2011, p. 5). Considering public data as a strategic resource, Australia's *Gov 2.0* initiatives are motivated by the potential of open data to "engage, innovate and create new public value" (AGIMO, 2010). To that end, Sayogo *et al.* (2014) explored the ethical dimension of open data and identified the motivations behind "smart disclosure". They defined smart disclosure as a government-led instrument that makes public and private data available in a timely manner to facilitate decision-making. They argued that such practice help consumers in making informed and better choices toward sustainable consumption. Accordingly, open data initiatives remain a high policy priority (Stott, 2014). The premise being that reuse and redistribution of public data has the potential to improve SME innovation performance as it can enhance efficacy of open data by improving market insights (Anderson, 2006) and, therein foster growth in niche markets (Fioretti, 2010).

5.3. *Improve efficiency, transparency and accountability*

As a mechanism to advance innovation, efficiency, transparency and accountability are not mutually exclusive factors, yet they are distinct motivators of engagement in open data initiatives (Davies & Perini, 2016; Dos Santos Brito *et al.*, 2014). The interactions between these factors have become fundamental to open data movement. The premise being that lack of transparency can leave the marginalized powerless and allow those in power to practice with limited regard for accountability, ultimately leading to inefficient market conditions (Fox, 2007; Joshi, 2013). Open access, open science, open education and open innovation initiatives of Canada's International Development Research Centre (IDRC) is a testament of how open data can make it possible for diverse groups of communities to come together, bringing transparency in process (Smith & Reilly, 2013a, 2013b). Yet, transparency is not sufficient to foster a culture of accountability (Kuriyan *et al.*, 2011). In considering Heald's (2006) upward, downward and horizontal transparency, open data promotes downward and horizontal transparency by allowing external actors (i.e., citizens, auditors) to monitor the internal governance networks (Fung, Graham & Weil, 2007).

In this context, transparency is often said to have the potential to enable new accountability networks affecting efficient and effective operations. For example, open data in journalism (Fink & Anderson, 2015; Martinisi, 2013) and open access to spending data (Maguire, 2011; Worthy, 2013) has held governments accountable for their actions. Further, social media has enabled and stimulated users to create new modes of social engagement (Pan *et al.*, 2007), fostering new motivation to increase transparency and efficiency through improvements in accountability (National Research Council, 2009; Pollock, 2009). It is then reasonable to promote open data as a strategic resource (Bates, 2014) for efficient functioning of markets (Davies & Perini, 2016). Examples of the use of open data to improve efficiency are prominent in practice. For example, Extractive Industries Transparency Initiative (EITI International Secretariat, 2012) has developed standardized tools for collection and publishing of data in a structured way. The standardized classification and reporting has enabled monitoring of contracts and government revenues, allowing for comparative analysis across nations and reporting periods. Likewise, OpenCorporates.com initiative in the EU has been actively promoting the use of open data to enhance corporate transparency and accountability in an effort to reduce market inefficiencies resulting from asymmetric information (Open Corporates, 2012).

5.4. *Promote inclusion through increase in social value*

Bentley & Chib (2016) posit that open development enabled by open data initiatives can "enable greater inclusion of poor and marginalized people and perspectives" (p. 3) by redistributing power, therein promoting a new way to collaborate. The premise is that with open data individuals and groups can build their own expositions of otherwise asymmetric information (Davies, 2010) across public and private data (Gigler *et al.*, 2014).

At this juncture, it is important to distinguish between open data initiatives currently adopted in practice which are "directed at" the marginalized groups and those that "directly engage" marginalized group in economic development activities (Rumbul, 2015). In the initiatives which are directed at the marginal groups, efforts are made to engage intermediaries (Chattapadhyay, 2014; Davies, 2010; Dumpawar, 2015) and enable groups to provide input into policy matters (Custer, 2012) through participatory community-based media and workshops (Canares, Marcial & Narca, 2016; De Boer *et al.*, 2012). On the other hand, initiatives that directly engage marginalized groups provide an active medium of exchange allowing citizens to act as agents empowered to shape policies in support of their social needs (Powell, Davies & Taylor, 2012). From a social and economic perspective, it can be argued that open data allows individuals and groups to ascertain and claim entitlements while reducing the workload of front-line departmental staff (Ubaldi, 2013).

6. Impediments in Open Data

Despite high expectations promoted in literature the impact of open data initiatives are yet to be realized (Bentley & Chib, 2016; Huijboom & Van den Broek, 2011). In fact, apart from the US and the UK, most other nations with open data initiatives do not have a mechanism to evaluate its value (Davies & Perini, 2016). To justify open data policies, governments tend to rely on previously published case studies with little to no recognition of contextual differences and implications (Huijboom & Van den Broek, 2011). For instance, Huijboom and Van den Broek (2011) in their survey of open government strategies across the US, the UK, Australia, Denmark and Spain found European Commission's PIRA (2000) and MEPSIR (2006) are the most cited studies in order to justify open data strategies. Likewise, many governments acknowledge that open data offers significant benefits (Open Government Partnership, 2017), yet the precise outcomes at organizational and regional levels are unclear (National Research Council, 2009). There seems even less evidence of social, technical and political impact of open data strategies (Hossain, Dwivedi & Rana, 2016; Huijboom & Van de Broek, 2011).

6.1. *Open data and socio-political risks*

A common value identified in most operational perspectives is that opening data builds trust between stakeholders (Meijer, Conradie & Choenni, 2014; O'Hara, 2012). However, research on transparency and trust in the context of open data has found mixed results. While some studies identified that open data improves trust in public agencies due to perception of increased control, others found that it decreases trust as failures become visible (Curtin & Meijer, 2006). It is also argued that open data may encourage politicians to engage in corrupt and misleading practices, obscuring performance toward economic development and blurring the boundaries between political leadership and administrative agenda of value-creation (Gurstein, 2011; Lessig, 2009).

Gurstein (2011) reasoned that access to data does not necessarily translate into social value and may in fact lead to "data divide" between those who are well-resourced and those who are not. Likewise while literature has put emphasis on the role of intermediaries in transferring the benefits of open data from the source to citizens (Chattapadhyay, 2014; Dumpawar, 2015; Van Schalkwyk, Willmers & McNaughton, 2015), it is argued that they may be biased toward securing self-interests (Moss & Coleman, 2014). Benjamin, Bhuvaneswari & Rayan (2007) in examining the impact of open data of land records in Bangladesh captured an example of such self-interest in the misuse of open data. They found that well-equipped citizens and organizations began to gain control of land from marginalized and under-resourced citizens. The authors found evidence of the "empowered", further marginalizing the marginalized by exploiting opportunities, mistakes and gaps in open data to gain control of land titles, often accentuated by a tendency to engage in corruption practices with intermediaries to the transactions. It can be argued that while open data is available to all, access to resources to translate data into information and insights may restrict the utility of open data, shifting power and fortune to those with skills and means to exploit the data for self-interest. The challenge in application of open data then rests in managing misuse and equitable opportunities to exploit the possibilities.

6.2. *Barriers to open data adoption*

It may be obvious that adoption, use and management of open data involves multiple stakeholders (i.e., policy makers, firms, intermediaries, research centers and society) at various stages of innovation (i.e., from conceptualization to commercialization). Integrating the needs, wants and desires of multiple stakeholders will be an ongoing challenge (Lassinantti, Bergvall-Kåreborn & Ståhlbröst, 2014). Yet, some progress has been made toward identifying barriers of adopting open data strategies (Hossain, Dwivedi & Rana, 2016; Ubaldi, 2013). While the "hard" (technical, legal and financial) limitations tend to affect utility of data, a number of barriers and challenges are associated with "soft" or behavioral aspects of engagement (Huijboom & van den Broek, 2011; Ubaldi, 2013).

Zuiderwijk *et al.* (2012) identified 118 social and technical challenges across 10 categories capturing the aspects of access to data, its utility and management. The categorical grouping of barriers has been a trend in literature. For instance, Sayogo & Pardo (2013) grouped the impediments into technological, social, organizational and economical, legal and policy, local context and specificity. The inclusion of local context as a barrier emphasizes the role of political and ecological impediments affecting the operating conditions. More recently, Barry & Bannister (2014) extended the local context to include cultural and administrative barriers. Cultural challenges refer to social pressures and consumer activism in favor of ethical behavior. Sayogo *et al.*, (2014) posited that a vital impediment to application of open data is the temporal nature of its utility. Timing of data release is critical to decision-making however, current administrative burden, uncertainty and costs associated with open data impede collection and credible utility of data (Sayogo *et al.*, 2014). Likewise, Zotti & La Mantia (2014) posited that information overload (volume of data), speed of information exchange (velocity), nature and extent of information (variety), and credibility of data (veracity) collectively capture the major barriers and challenges in management of open data. In Sections 5.2.1–5.2.4, we present an overview of the barriers and challenges found

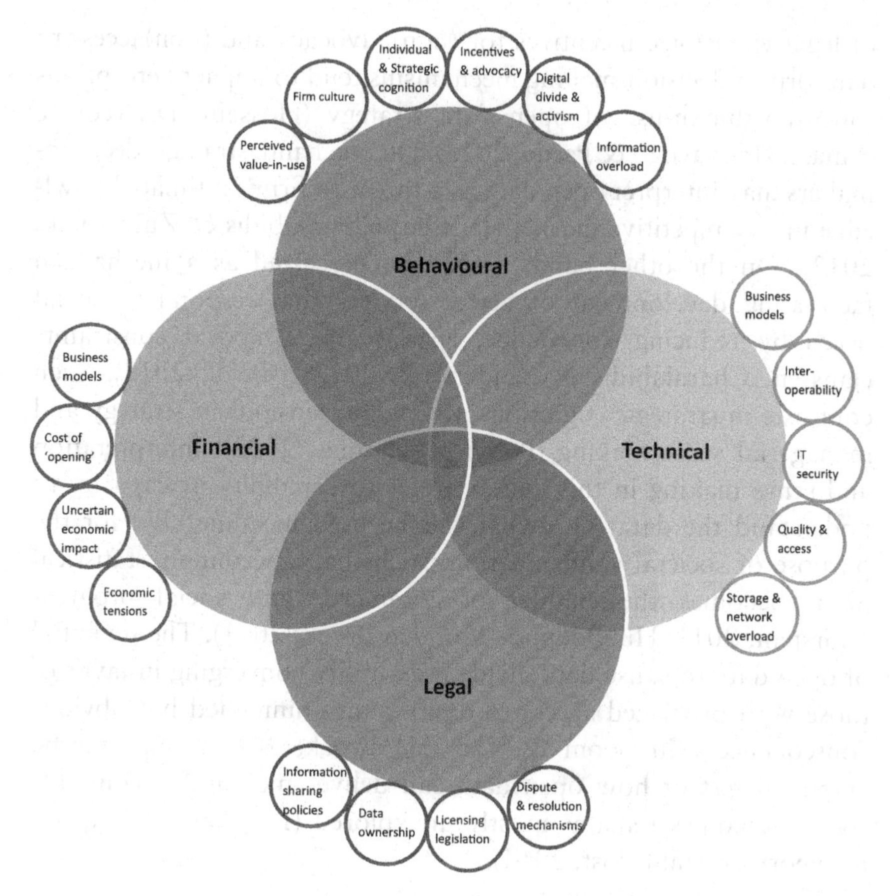

Fig. 2. Barriers and challenges in open data management

in current literature, grouped into behavioral, technical, legal and financial categories (see Fig. 2).

6.2.1. *Behavioral barriers and challenges*

Given its recent proliferation in practice and science, it is a reasonable assumption that most people do not understand the open data phenomenon and its functions. At the individual level, the issues relate to engagement and use of open data instruments, including awareness, knowledge and motivation to engage. Perceived threats of legal violations, incentives for (non)advocacy and (non)access to data-driven decision-making mechanisms tend to impact perceptions toward value-in-use of open data strategy (Hossain, Dwivedi & Rana, 2016; Sayogo & Pardo, 2013). On one hand, strategic decision-makers may interpret open data as a threat to firm's intimate knowledge in a competitive market (Janssen, Charalabidis & Zuiderwijk, 2012). On the other hand, it may be perceived as a mechanism facilitating development of better services that respond to social needs by reducing knowledge, spatial and temporal constraints (Janssen, Charalabidis & Zuiderwijk, 2012; Ubaldi, 2013). Such contrasts in strategic cognitions accentuate tensions in strategy and managerial sense making (Porac & Thomas, 2002). Interpretation and sense making in this context refer to capability or capacity to understand the data (i.e., what can be used meaningfully) for the purpose of societal value creation, such that a seemingly statistical input becomes the enabler of commerce and social progress (Gurstein, 2011; Huijboom & Van den Broek, 2011). The potential of open data to cause data divide (i.e., utility converging in favor of those with privileged access to data) is an unintended but obvious consequence (Huijboom & Van den Broek, 2011). Importantly, empirical test of how open data can deliver measurable equitable social outcomes remains mostly unexplored (Lassinantti, Bergvall-Kåreborn & Ståhlbröst, 2014).

6.2.2. *Technical barriers and challenges*

An important aspect for firms in embracing open data strategy is that it needs to be matched with related computing, statistical and technological skills (Conradie & Choenni, 2014). Studies exploring technical barriers have highlighted the challenges related to data quality, access to data, data credibility and interoperability (Huijboom & Van den Broek, 2011; Janssen, Charalabidis & Zuiderwijk, 2012). Interoperability of data is important for commercial applications (McAfee, Brynjolfsson & Davenport, 2012), yet it remains a major challenge (Whitmore, 2014). The process of standardizing data such that it allows effective reuse is complex

(Estermann, 2014) and requires significant commitment of resources. Currently, there are no universally accepted norms of open data and this remains the main hurdle in the utility of open data for managerial sense making. Thus, government and firms that are forced to embrace data under social and political pressures tend to release unstructured datasets, limiting the connections that can be established with other datasets to form meaningful patterns (Davies, Perini & Alonso, 2013). The lack of interoperable standards is a serious concern (Huijboom & Van Den Broek, 2011). Agencies tend to use a range of platforms and formats, which are often not compatible for data sharing. Data is thus often scattered, non-indexed, deposited in a non-open format or is in native language (Fleisher, 2008; Gurstein, 2011) and be laden with inconsistencies in terminology and structure (Linders, 2013). Further, storage and processing ability of large volume of data across current network and IT infrastructures remain as an ongoing trial (Fleisher, 2008; Zotti & La Mantia, 2014). There is a need for appropriate business models, technological infrastructure and know-how of converting data into insights (Zeleti, Ojo & Curry, 2014). Zuiderwijk & Janssen (2014) posited that to allow for appropriate interpretation of datasets it is essential for agencies to engage in standardized documentation and ensure credibility of data.

6.2.3. *Legal barriers and challenges*

Notably, open data is a global concept. The threats to privacy, confidentiality and security compromise the engagement and utility of open data. Managers argue that presence of suitable information policies governing distribution and use of data (i.e., copyright, licensing and reuse of contracts) could motivate them to employ open data strategies (Sayogo *et al.*, 2014). However, data collected by an agency has an initial purpose and this agency tends to limit its reuse (Hossain, Dwivedi & Rana, 2016). Concerns on who owns the data in an open environment (Conradie & Choenni, 2014), who pays for the costs of its production (Boulton *et al.*, 2011) remain the barriers to adoption in private and public settings (Sayogo *et al.*, 2014).

From a governance perspective, it is still unclear what legal ramifications could arise from misuse of open data (Dulong de Rosnay & Janssen, 2014; Kulk & Van Loenen, 2012) and how current legal frameworks could meet the demands for recourse on a global scale (Boulton, 2014). Is it the creator or the distributor that is liable for erroneous data? (Dulong de Rosnay & Janssen, 2014) and is open data another element of corporate social responsibility? (Boulton *et al.*, 2011). Some studies exploring the legal basis of open data policies aim to identify potential gaps in current legal frameworks (see Dulong de Rosnay & Janssen, 2014; Janssen, Charalabidis & Zuiderwijk, 2012; Kassen, 2013; Thurston, 2012). These studies integrate the notion of Right to Information (RTI) with open data models in practice — open government, open innovation and open science (Janssen, 2011). Tensions in such frameworks capture the legal aspects of data ownership and economic aspects of who pays and who benefits.

6.2.4. *Financial barriers and challenges*

Data-driven innovations require significant financial commitment from agencies in maintaining the efficacy of infrastructures. The cost of "opening" is thus a well-acknowledged barrier in implementing open data strategy (Huijboom & Van den Broek, 2011; Ubaldi, 2013). Nevertheless, to maintain financial feasibility, agencies engage in various modes of exchanges including charging for access or pay-per-use models (Hossain, Dwivedi & Rana, 2016). In this regard, paying for open data in an already low-quality and interoperability environment may seem counter-productive for the receiving agencies. Data is seen as a commodity by many private and public organizations (Ubaldi, 2013) and, thus barriers may emerge where agencies perceive open data as a threat to revenue (Hossain, Dwivedi & Rana, 2016). The debate on efficacy of open data is often rooted in societal value creation. The tension is then presented as being asymmetric: between the ethics of knowledge sharing and the politics of intellectual property security. In a dynamic and fast-changing world amid times of austerity and regulatory consolidation, complex problems could no longer be left to the invisible hand, rather collaborative partnerships in knowledge sharing leading to meaningful interpretations is needed (Robinson *et al.*, 2009; Ubaldi, 2013).

7. Toward an Open Data Culture

In practice, open data applications designed to suit semantic web are enhancing operability, interconnecting data from otherwise distributed networks using emerging technologies (Hendler *et al.*, 2012). These logic-driven extensions of the World Wide Web enhance capability by allowing computers to collect and collate standardized information across networks with little or no human intervention (Bizer *et al.*, 2008). In this regard, open data allow stakeholders to collaborate online as never before (De Souza *et al.*, 2015). It is then reasonable to deduce that open data can allow for manifestation of horizontal and vertical integration of creativity, promoting innovations in products and services for collective well-being. Progressively, more governments around the world are leveraging the positive externalities of open data by embracing "openness", prompting reconfigurations in policy and business models (European Commission, 2017b; Klímek *et al.*, 2017). Open government initiatives enable citizens and government to share a common picture of the intelligence that drives decisions across the nation by bringing public data to the web (Huber, Rentocchini & Wainwright, 2016). Not only does open data have the ability to promote transparency, accountability, governance at the organizational level and citizen engagement, it can enhance the richness, quality and accuracy of organization's own assets (OECD, 2016).

It is clear that open data quality characteristics include some parameters that are novel to the general corpus of data quality work. To operationalize and measure the quality of open data, there must be a way for an assessment to take place (Henke *et al.*, 2016). For instance, the UK government uses an algorithm based on the Five Stars of Linked Open Data to assess datasets when they are added to data.gov.uk, and weekly thereafter to reflect the fact that datasets are not absolute (Walker, Frank & Thompson, 2015). On the flipside, employment of open data strategy has been integrated as a performance indicator into the operational policies of many government departments and research centers, mostly due to its growing expectation as a prerequisite for securing public funds (Childs *et al.*, 2014). Likewise, as a measure of open science, some academic journals require authors to agree to share experimental data upon request (Andreoli-Versbach & Mueller-Langer, 2014) and some enforce a mandated release of research data (Tananbaum, 2008). The principle premise being that enforcing release of data will facilitate a cultural shift toward voluntary sharing of data (Hossain, Dwivedi & Rana, 2016). In this hope, many governments around the world have taken the lead to develop domain-specific digital collections of public data (Huijboom & Van den Broek, 2011), which are accessible for test and trials by all motivated and able members of the society (Hester, 2014). The motivation of national interest has been the central tenet of international open data development and this has extended to include open science. Open science projects in this context focus on innovation in the way research is conducted: in how scientists collaborate and share knowledge with the wider world (European Commission, 2017b). A key measure of the worth of research is the impact it has or, to put it in another way, the difference it is making both within the academic community and beyond (Henke *et al.*, 2016). Open science accelerates and coordinates research outcomes by facilitating global reach of scientists while vastly increasing the public accessibility of scientific knowledge across spatial and disciplinary boundaries. In recent years, funding bodies in Europe (Salmelin, 2015) and other parts of the world such as the United Kingdom and the United States (Centre for open data enterprise, 2016) have placed increasing emphasis on monitoring the potential and actual impact of the research projects they fund in the area of open data.

At the EU level for instance, Finland has been championing the culture of openness through open government and open science projects (Karvonen, 2017). Led by the Ministry of Finance, information resources on geo-data, climate, sea, transport, finance, statistics and culture have been opened in Finland since 2013, with integration at local, regional and national administrative levels. The Finnish open data projects have centered around four key elements of (1) digitization of public data, (2) promotion of standardized, machine-readable, free public information, (3) common model for production, service and management of information resources, and (4) enhancement of cross-sectoral data and information skills.

Similar to Finland in Europe, in the Oceania region Australian government has been leading charge on opening up regional and national data. Among many promises made in the Prime Minister's open data statement (Australian Government, 2015), much progress has been made toward making non-sensitive public data open by default across all government agencies, building partnerships with public, private and research centers and ensuring all new systems support discoverability, interoperability and accessibility in a cost-effective manner. Since 2013, the national and state governments in Australia have conducted extensive surveys and developed open data programs to include open data portals to promote use and reuse of spatial, statistical, sensor, cultural, research and science data. The developments in open science projects have focused on research data storage infrastructure, data catalogues, indexing service, identifier reference framework and information network. A key differentiation of the Australian movement toward open data culture is the integration of public and private data through community initiatives such as AustLII (legal records), GovHack, GovPond (aggregated government data) and OKFNau (discipline specific records). Now a single data portal (i.e., data.gov.au) provides easy access to open data in Australia and facilitates appeal process via public request functionality where a user is denied access to non-sensitive public data. Public data in this regard is a strategic resource and includes all data collected by government entities for any purposes including government administration, research or service delivery and, increasingly health and medical trial data.

An era of open information in healthcare is now underway. Progress has been evident in clinical research as pharmaceutical companies and other health organizations have aggregated years of research and development into open data portals. This has enabled a higher level of transparency and accountability to patients. Many governments in the EU are now mobilizing decades of stored data into usable and searchable open data in the public health sector. Because of such movements, healthcare stakeholders on an international scale have now have access to promising new threads of knowledge (Habl *et al.*, 2016). Communities are realizing the benefit of open data concept as sharing of medical information improves engagement toward health and well-being, and enhances the quality of patient care (Karvonen, 2017; Osterreich, 2016). Openness is perceived as the antidote that can counteract the tendency of technology enactment to reproduce existing rules, routines, norms and power relations (Hossain, Dwivedi & Rana, 2016). However, this perceived premise can only be fulfilled if openness changes the nature of relationships in a positive way and enables them to link across organizational boundaries and functions (European Commission, 2017a, 2017b).

8. The End of a Beginning

At national, regional, local and organizational levels, the capacity to remain competitive in the digital economy will remain contingent upon the ability to harness the value of data. Big data is growing exponentially and so too is the potential value of this data. Arguably, data is the next "oil". In the years to come, the entity that controls over production, service and management of data will maintain control over those who require it to function. However, open data publishing, linking through smart disclosure and sharing standardized data can create opportunities that neither government nor business can disparately envisage. Opening non-sensitive data has shown to stimulate innovation, promote start-ups, empower societies and deliver real sustainability outcomes.

Yet, powerful as open data's applications can be, valid concerns remain on how the benefits of open data can actually be maintained (Hossain, Dwivedi & Rana, 2016) and whether it can be trusted (Meijer, Conradie & Choenni, 2014). Even if an entity uses a third party and gets assurances of anonymity and confidentiality, there may always be a risk that individual identity might be revealed and that competitors could see how well or poorly it was doing (OECD, 2016). This is of a particular concern, especially in areas such as education, governance and healthcare sectors where reliance on performance and outcomes is often the foundation of engagement and funding strategy. At an individual level, the debate on open data focuses on privacy, security and skillset concerns (Hossain, Dwivedi & Rana, 2016). The tension is sometimes presented as being asymmetric: between the ethics of privacy and the politics of security, both affected by individual and organizational behavioral challenges. Thus, development of decision-making skills has been a common theme across most open data projects. It is essential to foster technical data science and information management skills, but in the age of digital data, it is equally critical to enhance civic skills of each individual to allow for critical assessment of data source and effective management of one's own information.

Open data can have a huge influence on the type of education that adults in general receive within organizations (Manyika *et al.*, 2013). Simply because organizations now have access to so many volumes of data, it is a step easier to train workers and to fully understand what kind of developmental needs they have within the entity itself (Henke *et al.*, 2016). Human resource managers can have access to much needed information across multiple divisions in an organization and across organizations, in order to train workers (Zang & Ye, 2015). This can impact productivity at the firm, and regional and national-level training can be developed to suit needs of specialization without constraints of organizational boundaries. However, managers face increased challenges in facilitating adoption and engagement of open data strategies. Every decision that is undertaken by organizations is underpinned by measurable and non-measurable information (Hossain, Dwivedi & Rana, 2016). Behavioral challenges arise from the realization that data is a strategic resource which needs to be protected from misuse and confidentiality concerns (Huber, Rentocchini & Wainwright, 2016).

Nevertheless, open data brings about a shift in culture as entities escape from an orientation toward procedural adherence and refocus on residual risk (outcomes). Additionally by helping launch new businesses, optimizing existing companies' operations, improving the climate for foreign investment (especially in the European Commission context), and bringing economic benefit to consumers of products and services (Reynolds, 2017), open data can create value both at the government level and the organizational level. Open data creates value in the private and public sectors in two basic ways: it provides the raw material for innovative new data-driven businesses, and it helps existing companies optimize their business by operating more efficiently and reaching customers more effectively (Australian Government, 2017). These can be as simple as development of mobile applications to enhance serviceability or as ambitious as major companies that are driven by weather, agriculture, geospatial and real estate data (OECD, 2016). While some regions have enjoyed a long tradition of relatively healthy research infrastructure (i.e., Europe) (Ramjoue, 2015), others are fostering the wave of digitization through open government, open science and open innovation initiatives at community, regional and national levels. And, this is not the end, nor is it the beginning of an end. Rather, it may be the end of the beginning of an open data phenomenon. Cliques (or groups) of open data proponents have indeed begun shifting the culture and the way of life as some used to know. Through the work of the European Commission and select other national entities (Open Government Partnership, 2017), people in general understand that for societal and economic progress there is a need to increase the level of sharing between employees, customers, organizations, research centers and governments. These early phases of the data revolution are important foundations for the future. Progress now depends on interconnected constellations of accessible, transparent, transferrable, free and trusted digital data.

References

AGIMO (2010). *Declaration of Open Government.* Retrieved from https://www.finance.gov.au/blog/2010/07/16/declaration-open-government.

Alexopoulos, C., Spiliotopoulou, L. & Charalabidis, Y. (2013). Open data movement in Greece: A case study on open government data sources. In *Proceedings of the 17th Panhellenic Conference on Informatics,* Thessaloniki, Greece, 19–21 September 2013, ACM, New York. pp. 279–286.

Anderson, C. (2006). *The Long Tail: Why the Future of Business Is Selling Less of More.* New York, NY: Hyperion.

Andreoli-Versbach, P. & Mueller-Langer, F. (2014). Open access to data: An ideal professed but not practised. *Research Policy, 43*(9), 1621–1633.

Australian Government (2015). *Public Data Policy Statement.* Retrieved from https://www.pmc.gov.au/sites/default/files/publications/aust_govt_public_data_policy_statement_1.pdf. Accessed 19 December 2017.

Australian Government (2017). *Open Data.* Retrieved from https://www.pmc.gov.au/public-data/open-data. Accessed 19 December 2017.

Bates, J. (2014). The strategic importance of information policy for the contemporary neoliberal state: The case of Open Government data in the United Kingdom. *Government Information Quarterly, 31*(3), 388–395.

Barry, E. & Bannister, F. (2014). Barriers to open data release: A view from the top. *Information Polity, 19*(1,2), 129–152.

Baue, B. & Murninghan, M. (2011). The accountability web: Weaving corporate accountability and interactive technology. *The Journal of Corporate Citizenship,* (41), 27.

Behkamal, B., Kahani, M., Bagheri, E. & Jeremic, Z. (2014). A metrics-driven approach for quality assessment of linked open data. *Journal of Theoretical and Applied Electronic Commerce Research, 9*(2), 64–79.

Benjamin, S., Bhuvaneswari, R. & Rajan, P. (2007). Bhoomi: 'E-governance', or, an anti-politics machine necessary to globalize Bangalore? *CASUM-m Working Paper.* Retrieved from https://casumm.files.wordpress.com/2008/09/bhoomi-e-governance.pdf. Accessed 19 December 2017.

Bentley, C. & Chib, A. (2016). The impact of open development initiatives in lower- and middle income countries: A review of the literature. *The Electronic Journal of Information Systems in Developing Countries,* 74.

Bertot, J., Gorham, U., Jaeger, P., Sarin, L. & Choi, H. (2014). Big data, open government and e-government: Issues, policies and recommendations. *Information Polity, 19*(1,2), 5–16.

Bichard, J. & Knight, G. (2012). Improving public services through open data: Public toilets. *Municipal Engineer, 165*(ME3), 157–165.

Bizer, C., Heath, T., Idehen, K. & Berners-Lee, T. (2008). Linked data on the web (LDOW2008). In *Proceedings of the 17th International Conference on World Wide Web*, Beijing, China, 21–25 April 2008, ACM: New York. pp. 1265–1266.

Bogers, M., Zobel, A. K., Afuah, A., Almirall, E., Brunswicker, S., Dahlander, L., Frederiksen, L., Gawer, A., Gruber, M., Haefliger, S. & Hagedoorn, J. (2017). The open innovation research landscape: Established perspectives and emerging themes across different levels of analysis. *Industry and Innovation*, *24*(1), 8–40.

Borzacchiello, M. & Craglia, M. (2012). The impact on innovation of open access to spatial environmental information: A research strategy. *International Journal of Technology Management*, *60*(1,2), 114–129.

Boulton, G. (2014). The open data imperative. *Insights*, *27*(2).

Boulton, G., Rawlins, M., Vallance, P. & Walport, M. (2011). Science as a public enterprise: The case for open data. *The Lancet*, *377*(9778), 1633–1635.

Bugshan, H. (2015). Open innovation using Web 2.0 technologies. *Journal of Enterprise Information Management*, *28*(4), 595–607.

Canares, M., Marcial, D. & Narca, M. (2016). Enhancing citizen engagement with Open Government data. *The Journal of Community Informatics*, *12*(2).

Caragliu, A., Del Bo, C. & Nijkamp, P. (2011). Smart cities in Europe. *Journal of Urban Technology*, *18*(2), 65–82.

Carrigan, M. & Attalla, A. (2001). The myth of the ethical consumer — Do ethics matter in purchase behaviour? *Journal of Consumer Marketing*, *18*(7), 560–578.

Carrara, W., Chan, W., Fische, S. & Steenbergen, E. (2015). Creating Value through Open Data: Study on the Impact of Re-use of Public Data Resources. *European Commission*. Retrieved from https://www.europeandataportal.eu/sites/default/files/edp_creating_value_through_open_data_0.pdf. Accessed 19 December 2017.

Castaldo, S., Perrini, F., Misani, N. & Tencati, A. (2009). The missing link between corporate social responsibility and consumer trust: The case of fair trade products. *Journal of Business Ethics*, *84*(1), 1–15.

Cerrillo-i-Martínez, A. (2012). Fundamental interests and open data for re-use. *International Journal of Law and Information Technology*, *20*(3), 203–222.

Chattapadhyay, S. (2014). Opening government data through mediation: Exploring the roles, practices and strategies of data intermediary organizations in India. *Open Data Research*. Retrieved from https://www.opendataresearch.org/sites/default/files/publications/sumandro_oddc_project_report_0.pdf. Accessed 19 December 2017.

Chesbrough, H. (2003). The logic of open innovation: Managing intellectual property. *California Management Review*, *45*(3), 33–58.

Chesbrough, H. & Bogers, M. (2014). Explicating Open Innovation: Clarifying an Emerging Paradigm for Understanding Innovation. In *New Frontiers in Open Innovation*, eds. H. Chesbrough, W. Vanhaverbeke & J. West, Oxford: Oxford University Press, p. 3.

Chief Secretary to the Treasury (2009). *Putting the Frontline First: Smarter Government*, London: HMGovernment. Retrieved from https://www.gov.uk/government/uploads/system/uploads/attachment_data/file/228889/7753.pdf. Accessed 19 December 2017.

Childs, S., McLeod, J., Lomas, E. & Cook, G. (2014). Opening research data: Issues and opportunities. *Records Management Journal*, *24*(2), 142–162.

Conradie, P. & Choenni, S. (2014). On the barriers for local government releasing open data. *Government Information Quarterly*, *31*, S10–S17.

Curtin, D. & Meijer, A. (2006). Does transparency strengthen legitimacy? *Information Polity*, *11*(2), 109–122.

Davenport, T. & Harris, J. (2007). *Competing on Analytics: The New Science of Winning*. Boston, MA: Harvard Business Press.

Davies, T. (2010). Open data, democracy and public sector reform. A look at open government data use from data.gov.uk. MSc thesis, Oxford Internet Institute.

Davies, T. & Perini, F. (2016). Researching the emerging impacts of open data: Revisiting the ODDC conceptual framework. *The Journal of Community Informatics*, *12*(2).

Davies, T., Perini, F. & Alonso, J. (2013). Researching the emerging impacts of open data in developing countries (ODDC). Retrieved from https://www.opendataresearch.org/sites/default/files/posts/Researching%20the%20emerging%20impacts%20of%20open%20data.pdf. Accessed 19 December 2017.

de Boer, V., de Leenheer, P., Bon, A., Baah Gyan, N., van Aart, C., Gueret, C. & Tuyp, W. (2012). Distributed Voice-en Web Interfaced Market Information System under Rural Conditions. In *Proceedings of 24th International Conference on Advanced Information Systems Engineering*, Gdansk, Poland, June 2012. Retrieved from http://www.few.vu.nl/~vbr240/publications/CAISE-2012_RadioMarche.pdf. Accessed 19 December 2017.

Doorey, D. (2011). The transparent supply chain: From resistance to implementation at Nike and Levi-Strauss. *Journal of Business Ethics*, *103*(4), 587–603.

Dos Santos Brito, K., da Silva Costa, M., Garcia, V. & de Lemos Meira, S. (2014). Brazilian government open data: Implementation, challenges, and potential opportunities. In *Proceedings of the 15th Annual International Conference on Digital Government Research*, Aguascalientes, Mexico, 18–21 June 2014, ACM, New York. pp. 11–16.

Dulong de Rosnay, M. & Janssen, K. (2014). Legal and institutional challenges for opening data across public sectors: Towards common policy solutions. *Journal of Theoretical and Applied Electronic Commerce Research*, *9*(3), 1–14.

Dumpawar, S. (2015). Open government data intermediaries: Mediating data to drive changes in the built environment, PhD thesis, Massachusetts Institute of Technology.

Dunleavy, P., Margetts, H., Bastow, S. & Tinkler, J. (2006). New public management is dead — Long live digital-era governance. *Journal of Public Administration Research and Theory*, *16*(3), 467–494.

Eisenhardt, K. (1989). Agency theory: An assessment and review. *Academy of Management Review*, *14*(1), 57–74.

EITI International Secretariat (2012). *Open EITI data*. Retrieved from https://eiti.org/explore-data-portal#eiti-summary-data. Accessed 19 December 2017.

Estermann, B. (2014). Diffusion of open data and crowdsourcing among heritage institutions: Results of a pilot survey in Switzerland. *Journal of Theoretical and Applied Electronic Commerce Research*, 9(3), 15–31.

European Commission (2003). *Directive 2003/98/EC of the European Parliament and of the council of 17 November 2003 on the re-use of public sector information*. Retrieved from https://ec.europa.eu/digital-single-market/overview-2003-psi-directive. Accessed 19 December 2017.

European Commission (2017a). *Horizon 2020 work programme 2016–2017*. Retrieved from http://ec.europa.eu/research/participants/data/ref/h2020/wp/2016_2017/main/h2020-wp1617-intro_en.pdf. Accessed 19 December 2017.

European Commission (2017b). *Enter the data economy: EU policies for a thriving data ecosystem*. European Political Strategy Centre. Retrieved from https://ec.europa.eu/epsc/publications/strategic-notes/enter-data-economy_en. Accessed 19 December 2017.

Fayyad, U., Piatetsky-Shapiro, G. & Smyth, P. (1996). From data mining to knowledge discovery in databases. *AI magazine*, *17*(3), 37.

Fink, K. & Anderson, C. (2015). Data journalism in the United States: Beyond the "usual suspects". *Journalism Studies*, *16*(4), 467–481.

Fioretti, M. (2010). Open data, open society: A research project about openness of public data in EU local administration. *Laboratory of Economics and Management of Scuola Superiore Sant'Anna*, Pisa.

Fleisher, C. (2008). Using open source data in developing competitive and marketing intelligence. *European Journal of Marketing*, *42*(7/8), 852–866.

Fox, J. (2007). The uncertain relationship between transparency and accountability. *Development in Practice*, *17*(4,5), 663–671.

Fung, A., Graham, M. & Weil, D. (2007). *Full Disclosure: The Perils and Promise of Transparency*. Cambridge: Cambridge University Press.

Fung, A. & Weil, D. (2010). Open Government and Open Society. In *Open Government: Collaboration, Transparency, and Participation in Practice*, eds. D. Lathrop & L. Ruma, Sebastopol, CA: O'Reilly Media. 105–113.

Garbett, A., Linehan, C., Kirman, B., Wardman, J. & Lawson, S. (2010). Using social media to drive public engagement with open data. *Digital Engagement*, *11*.

Gaudiano, P. (2017). The best approach to decision making combines data and managers' expertise, *Harvard Business Review*, 20 June. Retrieved from https://hbr.org/2017/06/the-best-approach-to-decision-making-combines-data-and-managers-expertise. Accessed 19 December 2017.

Gerunov, A. (2015). Open data: Policy and implementation in Bulgaria. *Munich Personal RePEc Archive*. Retrieved from https://mpra.ub.uni-muenchen.de/68799/1/MPRA_paper_68799.pdf. Accessed 19 December 2017.

Gigler, B., Custer, S., Bailur, S., Dodds, E., Asad, S. & Gagieva-Petrova, E. (2014). Closing the feedback loop: Can technology amplify citizen voices. *Closing the Feedback Loop*, 211.

Gurstein, M. (2011). Open data: Empowering the empowered or effective data use for everyone? *First Monday*, *16*(2).

Habl, C., Renner, A. T., Bobek, J. & Laschkolnig, A. (2016). Study on Big Data in Public Health, Telemedicine and Healthcare. Final Report.

Harrison, T. & Sayogo, D. (2013). Open budgets and open government: Beyond disclosure in pursuit of transparency, participation and accountability. In *Proceedings of the 14th Annual International Conference on Digital Government Research*, Quebec City, Canada, 17–20 June 2013, ACM: New York. pp. 235–244.

Hartung, C., Lerer, A., Anokwa, Y., Tseng, C., Brunette, W. & Borriello, G. (2010). Open data kit: Tools to build information services for developing regions. In *Proceedings of the 4th ACM/IEEE International Conference on Information and Communication Technologies and Development*, London, UK, December 2010, ACM: New York. p. 18.

Heald, D. (2006). Varieties of Transparency. In *Transparency: The Key to Better Governance?* eds. Hood, C. & Heald, D., Proceedings of the British Academy (135). Oxford: Oxford University Press for The British Academy. 25–43.

Hendler, J. & Berners-Lee, T. (2010). From the Semantic Web to social machines: A research challenge for AI on the World Wide Web. *Artificial Intelligence*, *174*(2), 156–161.

Hendler, J., Holm, J., Musialek, C. & Thomas, G. (2012). US government linked open data: Semantic. data. gov. *IEEE Intelligent Systems*, *27*(3), 25–31.

Henke, N., Bughin, J., Chui, M., Manyika, J., Saleh, T., Wiseman, B. & Sethupathy, G. (2016). The age of analytics: Competing in a data-driven world. *McKinsey Global Institute*, 4.

Hester, J. (2014). Closing the data gap: Creating an open data environment. *Radiation Physics and Chemistry*, *95*, 59–61.

Hippel, E. & Krogh, G. (2003). Open source software and the "private-collective" innovation model: Issues for organization science. *Organization Science*, *14*(2), 209–223.

Hossain, M., Dwivedi, Y. & Rana, N. (2016). State-of-the-art in open data research: Insights from existing literature and a research agenda. *Journal of Organizational Computing and Electronic Commerce*, *26*(1,2), 14–40.

Huber, F., Rentocchini, F. & Wainwright, T. (2016). Open Innovation: Revealing and Engagement in Open Data organizations (No. 2016–19). *SPRU-Science and Technology Policy Research*, University of Sussex.

Huijboom, N. & Van den Broek, T. (2011). Open data: An international comparison of strategies. *European Journal of Epractice*, *12*(1), 1–13.

Jackson, M. (2003). *Systems Thinking: Creative Holism for Managers*. Chichester, UK: Wiley.

Janssen, K. (2011). The influence of the PSI directive on open government data: An overview of recent developments. *Government Information Quarterly*, *28*(4), 446–456.

Janssen, K. (2012). Open government data and the right to information: Opportunities and obstacles. *The Journal of Community Informatics*, *8*(2).

Janssen, M., Charalabidis, Y. & Zuiderwijk, A. (2012). Benefits, adoption barriers and myths of open data and open government. *Information Systems Management*, *29*(4), 258–268.

Joshi, A. (2013). Do they work? Assessing the impact of transparency and accountability initiatives in service delivery. *Development Policy Review*, *31*(s1).

Kalampokis E., Tambouris E., Tarabanis K. (2011) Open Government Data: A Stage Model. In: Janssen M., Scholl H.J., Wimmer M.A., Tan Y. (eds) Electronic Government. EGOV 2011. Lecture Notes in Computer Science, vol 6846. Springer, Berlin, Heidelberg.

Karvonen, M. (2017). Open data initiatives in Finland. In *Proceedings at Nordic CIO conference*, Hanken, Helsinki, 31 March 2017. Retrieved from https://www.helsinki.fi/sites/default/files/atoms/files/minna_karvonen.pdf. Accessed 19 December 2017.

Kassen, M. (2013). A promising phenomenon of open data: A case study of the Chicago open data project. *Government Information Quarterly*, *30*(4), 508–513.

Klímek, J., Kučera, J., Nečaský, M. & Chlapek, D. (2017). Publication and usage of official Czech pension statistics Linked Open Data. *Web Semantics: Science, Services and Agents on the World Wide Web*. Retrieved from file://ntapprdfs01n01.rmit.internal/el5/e30415/Downloads/507-895-1-SM.pdf. Accessed 19 December 2017.

Krumholz, H. & Waldstreicher, J. (2016). The Yale Open Data Access (YODA) project — A mechanism for data sharing. *New England Journal of Medicine*, *375*(5), 403–405.

Kucera, J., Chlapek, D., Klímek, J. & Necaský, M. (2015). Methodologies and best practices for open data publication. In *Proceedings of 15th Annual International Workshop on. Databases, Texts, Specifications, and Objects* (DATESO), Nepřívěc u Sobotky, Jičín, Czech Republic, 14–16 April 2015. pp. 52–64.

Kulk, S. & Van Loenen, B. (2012). Brave new open data world. *International Journal of Spatial Data Infrastructures Research*, 7.

Kuriyan, R., Bailur, S. Gigler, S. & Ryul Park, K. (2011). Technologies for Transparency and Accountability: Implications for ICT Policy and Implementation. *World Bank, Open Development Technology Alliance*, Washington, D.C.

Lassinantti, J., Bergvall-Kåreborn, B. & Ståhlbröst, A. (2014). Shaping local open data initiatives: Politics and implications. *Journal of Theoretical and Applied Electronic Commerce Research*, *9*(2), 17–33.

Lathrop, D. & Ruma, L. (2010). *Open Government: Collaboration, Transparency, and Participation in Practice*. Sebastopol, CA: O'Reilly Media.

Laursen, K. & Salter, A. (2006). Open for innovation: The role of openness in explaining innovation performance among UK manufacturing firms. *Strategic Management Journal*, *27*(2), 131–150.

Lessig, L. (2009). *Against Transparency: The perils of openness in government*, The New Republic. Retrieved from https://newrepublic.com/article/70097/against-transparency. Accessed 19 December 2017.

Lin, X., Li, Y. & Wang, X. (2017). Social commerce research: Definition, research themes and the trends. *International Journal of Information Management*, *37*(3), 190–201.

Linders, D. (2013). Towards open development: Leveraging open data to improve the planning and coordination of international aid. *Government Information Quarterly*, *30*(4), 426–434.

Linders, D. & Wilson, S. (2011). What is open government? One year after the directive. In *Proceedings of the 12th Annual International Digital Government Research Conference: Digital Government Innovation in Challenging Times*, Washington, D.C., 12–15 June 2011, ACM: New York. pp. 262–271.

Lindman, J. (2014). Similarities of open data and open source: Impacts on business. *Journal of Theoretical and Applied Electronic Commerce Research*, *9*(3), 46–70.

Maier-Rabler, U. & Huber, S. (2012). "Open": The changing relation between citizens, public administration, and political authority. *JeDEM-eJournal of eDemocracy and Open Government*, *3*(2), 182–191.

Maguire, S. (2011). Can data deliver better government? *The Political Quarterly*, *82*(4), 522–525.

Manyika, J., Chui, M., Groves, P., Farrell, D., Van Kuiken, S. & Doshi, E. (2013). Open data: Unlocking innovation and performance with liquid information. *McKinsey Global Institute*, October. Retrieved from https://www.mckinsey.com/business-functions/digital-mckinsey/our-insights/open-data-unlocking-innovation-and-performance-with-liquid-information. Accessed 19 December 2017.

Martinisi, A. (2013). Data journalism and its role in open government. *Challenges, Solutions, Knowledge Models in e-governance*, 58.

McAfee, A., Brynjolfsson, E. & Davenport, T. (2012). Big data: The management revolution. *Harvard Business Review*, *90*(10), 60–68.

McGee, R. & Edwards, D. (2016). Introduction: Opening governance — Change, continuity and conceptual ambiguity. *IDS Bulletin*, *47*(1).

McGee, R., Edwards, D., Minkley, G., Pegus, C. M. & Brock, K. (2015). Making All Voices Count Research and Evidence Strategy, Brighton, Institute of Development Studies.

McLeod, J. (2012). Thoughts on the opportunities for records professionals of the open access, open data agenda. *Records Management Journal*, *22*(2), 92–97.

Meijer, R., Conradie, P. & Choenni, S. (2014). Reconciling contradictions of open data regarding transparency, privacy, security and trust. *Journal of Theoretical and Applied Electronic Commerce Research*, *9*(3), 32–44.

MEPSIR (2006). *Measuring European Public-Sector Information Resources, Final Report of Study on Exploitation of public sector information — Benchmarking of EU framework conditions*. European Commission: Brussels.

Moss, G. & Coleman, S. (2014). Deliberative manoeuvres in the digital darkness: e-democracy policy in the UK. *The British Journal of Politics and International Relations*, *16*(3), 410–427.

National Research Council (2009). *The Socioeconomic Effects of Public Sector Information on Digital Networks: Toward a Better Understanding of Different Access and Reuse Policies: Workshop Summary*. Washington DC: National Academies Press.

Noveck, B. (2009). *Wiki Government: How Technology Can Make Government Better, Democracy Stronger, and Citizens More Powerful*. Washington, DC: Brookings Institution Press.

Obama, B. (2012). *Digital Government. Building a 21st Century Platform to Better Serve the American People*. The White House Archives. Retrieved from https://obamawhitehouse.archives.gov/sites/default/files/omb/egov/digital-government/digital-government.html. Accessed 19 December 2017.

OECD (2016). *Compendium of Good Practices on the Publication and Reuse of Open Data for Anti-corruption across G20 Countries: Towards Data-driven Public Sector Integrity and Civic Auditing*. Retrieved from https://www.oecd.org/gov/digital-government/g20-oecd-compendium.pdf. Accessed 19 December 2017.

O'Hara, K. (2012). Transparency, open data and trust in government: Shaping the infosphere. In *Proceedings of the 4th Annual ACM Web Science Conference*, Evanston, IL, 22–24 June 2012, ACM, New York. pp. 223–232.

Open Corporates (2012). *The Closed World of Company Data: An examination of How Open Company Data is in Open Government Partnership Countries*. Retrieved from https://www.access-info.org/wp-content/uploads/Closed_World_Company_Data.pdf. Accessed 19 December 2017.

Open Government Partnership (2017). *Participants*. Retrieved from https://www.opengovpartnership.org/participants. Accessed 18 December 2017.

O'Reilly, T. (2011). Government as a Platform. *Innovations*, *6*(1), 13–40.

Pabón, G., Gutiérrez, C., Fernández, J. & Martínez Prieto, M. (2013). Linked Open Data technologies for publication of census microdata. *Journal of the Association for Information Science and Technology*, *64*(9), 1802–1814.

Pan, B., Hembrooke, H., Joachims, T., Lorigo, L., Gay, G. & Granka, L. (2007). In google we trust: Users' decisions on rank, position, and relevance. *Journal of Computer-Mediated Communication*, *12*(3), 801–823.

Peled, A. (2011). When transparency and collaboration collide: The USA open data program. *Journal of the Association for Information Science and Technology*, 62(11), 2085–2094.

Petrov, O., Gurin, J. & Manley, L. (2015). Open Data for Sustainable Development (No. Policy Note ICT01). *World Bank*. Retrieved from http://www.worldbank.org/en/topic/ict/brief/open-data-for-sustainable-development. Accessed 19 December 2017.

Phoensight (2016). *Open Data Supply: Enriching the Usability of Information.* Retrieved from https://www.pc.gov.au/__data/assets/pdf_file/0009/198765/sub002-data-access.pdf. Accessed 19 December 2017.

PIRA International (2000). Commercial exploitation of Europe's Public-Sector Information. Final Report. *European Commission*, Surrey, Pira International.

Pollock, R. (2009). The economics of public sector information. *Working Paper.* University of Cambridge. Retrieved from https://www.repository.cam.ac.uk/bitstream/handle/1810/229487/0920.pdf?sequence=2. Accessed 19 December 2017.

Porac, J. & Thomas, H. (2002). Managing cognition and strategy: Issues, trends and future directions. In *Handbook of Strategy and Management*, 165–181.

Powell, M., Davies, T. & Taylor, K. (2012). ICT for or against development? An introduction to the ongoing case of Web 3.0. *IKM Emergent Research Programme, European Association of Development Research and Training Institutes* (*EADI*). Retrieved from http://wiki.ikmemergent.net/files/1204-IKM-Working_Paper_16-WEB3-Mar_2012-2.pdf. Accessed 19 December 2017.

Provost, F. & Fawcett, T. (2013). *Data Science for Business: What You Need to Know about Data Mining and Data-analytic Thinking.* Sebastopol, CA: O'Reilly Media.

Ramjoué, C. (2015). Towards open science: The vision of the European Commission. *Information Services & Use*, *35*(3), 167–170.

Reynolds, F. (2017). Open Banking: A consumer perspective. *Barclays.* Retrieved from https://www.home.barclays/content/dam/barclayspublic/docs/Citizenship/Research/Open%20Banking%20A%20Consumer%20Perspective%20Faith%20Reynolds%20January%202017.pdf. Accessed 19 December 2017.

Roberts, D. & Piller, F. (2016). Finding the right role for social media in innovation. *MIT Sloan Management Review*, *57*(3), 41.

Robinson, D., Yu, H., Zeller, W. & Felten, E. (2009). Government data and the invisible hand. *Yale Journal of Law and Technology*, *11*(1), 4.

Rohunen, A., Markkula, J., Heikkila, M. & Heikkila, J. (2014). Open traffic data for future service innovation: Addressing the privacy challenges of driving data. *Journal of Theoretical and Applied Electronic Commerce Research*, *9*(3), 71–89.

Rothenberg, J. (2012). Towards a better supply and distribution process for open data: In case study international benchmark on open data and use of standards. *Forum Standaardisatie*.

Rumbul, R. (2015). Who benefits from civic technology? Demographic and public attitudes research into the users of civic technologies. *mySociety*. Retrieved from https://www.mysociety.org/research/who-benefits-fromcivic-technology. Accessed 19 December 2017.

Salmelin, B. (2015). *Open Innovation 2.0: Yearbook*. European Commission.

Sayogo, D. & Pardo, T. (2013). Exploring the determinants of scientific data sharing: Understanding the motivation to publish research data. *Government Information Quarterly*, *30*, 19–31.

Sayogo, D., Zhang, J., Pardo, T., Tayi, G., Hrdinova, J., Andersen, D. & Luna-Reyes, L. (2014). Going beyond open data: Challenges and motivations for smart disclosure in ethical consumption. *Journal of Theoretical and Applied Electronic Commerce Research*, *9*(2), 1–16.

Shadbolt, N., O'Hara, K., Berners-Lee, T., Gibbins, N., Glaser, H. & Hall, W. (2012). Linked open government data: Lessons from data.gov.uk. *IEEE Intelligent Systems*, *27*(3), 16–24.

Smith, M. & Reilly, K. (eds.) (2013a). *Open Development: Networked Innovations in International Development*. Cambridge, MA: MIT Press.

Smith, M. & Reilly, K. (2013b). The Emergence of Open Development in a Network Society. In *Open Development: Networked Innovations in International Development*, eds. M. L. Smith & K. M. A. Reilly, Cambridge, MA: MIT Press. 15–50.

Srimarga, I., Suhaemi, A., Narhetali, E., Wahyuni, I., Rendra, M., Firmansyah, S. & Heriyanto, W. (2014). *Open Data Initiative of Ministry of Finance on National Budget Transparency in Indonesia*. ODDC. Retrieved from https://idl-bnc-idrc.dspacedirect.org/bitstream/handle/10625/56293/IDL-56293.pdf?sequence=1. Accessed 19 December 2017.

Stott, A. (2014). *Open Data for Economic Growth*. Retrieved from http://www.worldbank.org/content/dam/Worldbank/document/Open-Data-for-Economic-Growth.pdf. Accessed 19 December 2017.

Sunstein, C. (2011). *Informing consumers through smart disclosure*. White House: Washington, D.C.

Tananbaum, G. (2008). Adventures in open data. *Learned Publishing*, *21*(2), 154–156.

Thurston, A. (2012). Trustworthy records and open data. *The Journal of Community Informatics*, *8*(2).

Tolbert, C. & Mossberger, K. (2006). The effects of e-government on trust and confidence in government. *Public Administration Review*, 66(3), 354–369.

Ubaldi, B. (2013). Open government data: Towards empirical analysis of open government data initiatives. *OECD Working Papers on Public Governance*, (22), 1.

Van Schalkwyk, F., Willmers, M. & McNaughton, M. (2016). Viscous open data: The roles of intermediaries in an open data ecosystem. *Information Technology for Development*, 22(1), 68–83.

Vickery, G. (2011). Review of recent studies on PSI re-use and related market developments. *Information Economics, Paris.*

Walker, J., Frank, M. & Thompson, N. (2015). User centred methods for measuring the value of open data. In *Proceedings at Open Data Research Symposium*, Canada, 27 May 2015. p. 31.

Web Foundation (2017). *Open Data Barometer 4th Edition — Global Report.* Retrieved from http://opendatabarometer.org/doc/4thEdition/ODB-4thEdition-GlobalReport.pdf. Accessed 19 December 2017.

Whitmore, A. (2014). Using open government data to predict war: A case study of data and systems challenges. *Government Information Quarterly*, *31*(4), 622–630.

Worthy, B. (2013). *David Cameron's Transparency Revolution? The Impact of Open Data in the UK.* Retrieved from https://papers.ssrn.com/sol3/papers.cfm?abstract_id=2361428. Accessed 19 December 2017.

Zang, S. & Ye, M. (2015). Human Resource Management in the Era of Big Data. *Journal of Human Resource and Sustainability Studies*, *3*(01), 41.

Zeleti, F., Ojo, A. & Curry, E. (2014). Emerging business models for the open data industry: characterization and analysis. In *Proceedings of the 15th Annual International Conference on Digital Government Research*, Aguascalientes, Mexico, 18–21 June 2014, ACM, New York. pp. 215–226.

Zotti, M. & La Mantia, C. (2014). Open data from earth observation: From big data to linked open data, through INSPIRE. *Journal of e-Learning and Knowledge Society*, *10*(2).

Zuiderwijk, A. & Janssen, M. (2014). Barriers and development directions for the publication and usage of open data: A socio-technical view. In *Open government*, New York, NY: Springer. 115–135.

Zuiderwijk, A., Janssen, M. & Choenni, S. (2012). Open data policies: Impediments and challenges. In *Proceedings of the 12th European Conference on e-government (ECEG 2012)*, Barcelona, Spain, 14–15 June 2012, Academic Conferences and Publishing International Limited. pp. 794–802.

Zuiderwijk, A., Helbig, N., Gil-García, J. & Janssen, M. (2014). Special Issue on Innovation through Open Data: Guest Editors' Introduction. *Journal of Theoretical and Applied Electronic Commerce Research*, 9(2), pp. 1–13.

Zuiderwijk, A., Janssen, M., Choenni, S., Meijer, R. & Alibaks, R. (2012). Socio-technical Impediments of Open Data. *Electronic Journal of e-Government*, 10(2).

Zuiderwijk, A., Janssen, M. & Davis, C. (2014). Innovation with open data: Essential elements of open data ecosystems. *Information Polity*, *19*(1,2), 17–33.

Zurada, J. & Karwowski, W. (2011). Knowledge discovery through experiential learning from business and other contemporary data sources: A review and reappraisal. *Information Systems Management*, *28*(3), 258–274.

Chapter 2

Stimulation of Open Data Ecosystems: Learnings from Theory and Practice

Slim Turki, Sébastien Martin and Samuel Renault

Luxembourg Institute of Science and Technology, Luxembourg

This chapter addresses the stimulation of open data ecosystems. Indeed, since the first data release, open data generated many expectations in the economic and social domains. Most of these benefits remain, however, limited and their sustainability might be challenged. Indeed, value creation from open data faces several challenges, among them are the risk of being too supply-driven, or the lack of incentives for the reuse. The idea of stimulating an ecosystem through the mean of innovative legal and financial frameworks is not entirely new. From an empirical point of view, we identify several meaningful initiatives undertaken by different countries and trying to bypass the obstacles faced by potential open data reusers to stimulate economic growth. For each initiative, the main lessons learnt are highlighted. From a theoretical point of view, the ecosystem perspective is widespread in open data research. In these systems, we find room for a new role, at least a transversal role, consisting in stimulating the ecosystem. This role is specific in that it implies to understand the configuration and the mechanisms of the ecosystems, and to define an influence strategy. We propose a framework able to guide a given stimulator in its analysis and related actions on its open data ecosystem. Then, we show in return how the practice could be enriched by this framework and how it could be instantiated and refined in a European program designed to foster the innovativeness of the open data ecosystems.

1. Introduction

Open data movement has mainly been a data provision movement. The release of open data is usually motivated by

1. government transparency (citizens' access to government data);
2. development of services by third parties (Foulonneau *et al.*, 2014a) for the benefit for citizens and companies (typically smart city approach) or;
3. development of new services that contribute to the economy (Martin *et al.*, 2013; Foulonneau *et al.*, 2014b).

In this chapter, we focus on the third dimension, as part of the review of experiences with public sector information (PSI) initiatives. After a first and euphoric phase, as there is still a lack of evidence of concrete economic value, some open data promoters have encountered delusion and wonder if the original hopes were overstated or if the somewhat meager results can be explained by failures in the ways data have been released (Magalhães *et al.*, 2016). There is an agreement on the fact that an ecosystem perspective, where open data is embedded in much larger issues can be a clue to success in reaching the goals set by open data promoters.

This chapter presents first a selection of success stories related to the development of local ecosystems. Each one bears meaningful insights on the ideas, implementation methods and assessment models already developed. Some initiatives are already trying to deal with the demand side of open data, at different scales and with different levers. Even if most of them are successful, not all are accurately documented and we still lack the basis of a methodology to influence an ecosystem. In consequence, we try then to amplify, generalize and partially model these insights in order to define the lineaments of a systematic influence process of the open data ecosystem.

In Martin, Turki & Renault (2017), we introduced the theoretical grounds of a stimulation function where we discussed the relevance to adapt the frameworks developed for other kinds of ecosystems or platforms, mostly from the field of strategic management, to open data ecosystems. This role entails for a given actor to be aware of the existence of its ecosystem and to develop a strategy to orchestrate the products of services generated and to manage or influence the relationships among the actors.

Existing literature on open data does not ignore the role of public bodies to influence the shape of the ecosystem, the nature of its products, the ways they are produced and the kinds of relationships that are required. However, there is a lack of a consistent framework to orchestrate and assess these strategic interventions. In consequence, we propose to integrate public actors in the framework previously defined.

Finally, we show how a reliable instrument of influence is public procurement, increasingly envisioned in a strategic perspective allowing to engage in capacity building for the modernization of the private sector.

This rest of this chapter is organized as follows: Section 2 presents some meaningful initiatives coping with the demand side of open data, open data incubators are the focus in Section 3, Section 4 summarizes the interest of an ecosystem perspective; Section 5 presents a model of open data stimulation and Section 6 explains how the BE-GOOD program intends to instantiate this stimulation function.

2. Insights from Prior Open Data Initiatives

The following cases were already introduced in Turki, Martin and Renault (2017).

2.1. *Singapore — Call-for-Collaboration (CFC)*

One interesting, and unique as far as we know, example to encourage the creation of valuable services from public data beyond hackathons is the Call-for-Collaboration (CFC) model set up in Singapore (Chan, 2013; Land authority of Singapore, 2015).

In terms of incentives for reuse, Singapore has put in place a dual strategy: like many public institutions (states, regions, etc.), Singapore organized hackathons that allowed to develop applications then referenced on its platform of open data. This can be seen as the first step in an open data reuse policy, for which many initiatives are currently standing.

Several considerations have led this government to explore other ways of value creation with open data. Among them are the limits faced by the model of hackathon (see Section 2.3) or the small size of the Singaporean population which can deter initial investments from the private sector. Moreover, according to this government, the innovative potential of hackathon was limited by the fact that the open data ecosystem suffers from a lack of innovativeness.

Aware of these limits (Martin *et al.*, 2013; Foulonneau *et al.*, 2014B), Singapore's government used an instrument previously existing and grounded in the stimulation of innovation.

There is indeed a long tradition of intervention on the course of the economic development, where the state intends to play a strategic role, which requires a much higher level of funding but allows to indicate precisely the kind of service the institution wishes to generate and especially to foster the growth of a given economic sector. This must be framed in a broader context of Singapore's constant adaptation to the new economic contexts. Chua (2011) lists building public–private partnerships (PPPs) among the levers that have allowed Singapore to take advantage of successive waves of innovation and to be among the most successful economies.

CFC instrument fits well with Singapore's development. Singapore's government has therefore used the CFC approach. CFC methodology was not conceived originally for the open data context but has been combined with the open data principles. Before the

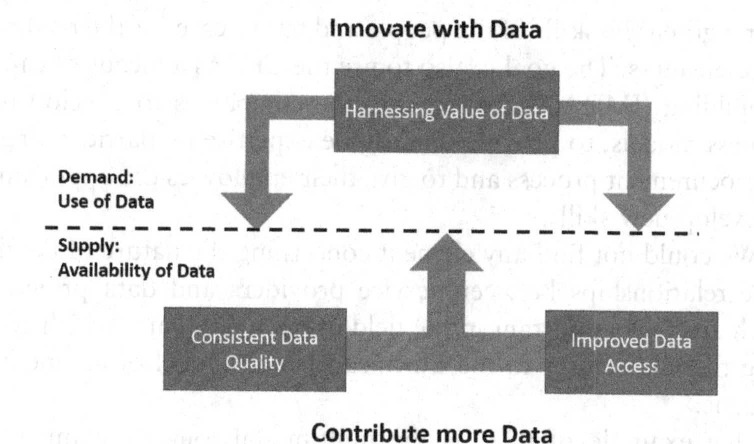

Fig. 1. Data-driven innovation strategy in Singapore, to create a sustainable data ecosystem to catalyze demand and drive supply of government data

reform of 2016, the Infocomm Development Authority of Singapore (IDA) was a public agency in charge of the "Intelligent Nation 2015" program, whose goal was to stimulate the growth of the IT sector in Singapore. One of the subprograms concerned the theme of e-government.

It is an original framework (for 2013) of a methodology combining open data with public intervention in economic development. It might seem self-evident for people accustomed with strong public intervention, but it deviates at least in part from several principles supported by some of the first open data evangelists, who thought that the role of public administration could stop after the provision of data (Fig. 1).

It encompasses a wide variety of topics and actors. In the frame of the CFC, even if it does not reach the degree of accuracy of a tender specification, the public agency details the functionalities of the products. This could be the first basis on a convenient strategy aiming to engage reusers in a long term and issue-driven collaboration.

Besides the provision of new products and services, one objective of the kind of PPP that is the CFC is clearly to build an ecosystem through the release of datasets, the local firms and people (start-ups), to strengthen the skills of the people and to increase for the readiness of the business. The goal is also to use the CFC as a means of capacity building (IMDA, 2016) to help the companies to develop new business models, to provide them some expertise by participating in the procurement process and to give their employees the opportunity to develop new skills.

We could not find any element concerning the nature of the dialogue relationships between service providers and data providers, which are very important in the field of open data, and which could bring the CFC closer to the European Union model of competitive dialogue.

One example of CFC is the multimodal journey planner for Singapore (LTA, 2015), launched in 2015 under the auspices of the Land Transport Authority agency. The call for collaboration lists in a very broad way the main features of the service and gives some minimal constraints, for example, language settings, and kinds of alerts available for final users. Proposals must use at least one government dataset from any of these areas: business planning, conservation and environment, contextual content, connecting people, optimized mobility and healthy lifestyle. The guidelines document draws the evaluation criteria very broadly too and even with some vagueness. The innovativeness criterion is mentioned, but without explaining what the organizer meant there, or what aspects will be retained for evaluation. The submission date is set two months after the publication of the call, which seems rather few. The CFC got few proposals. We have identified only four applications. This is one of the variables on which the organizer has only indirect levers and depends on the maturity of other components of the ecosystem.

One question concerns the impulsions necessary to reach homeostasis in the ecosystem. This could be the role of incubators and open data platforms, but it seems to be a rather long-term goal. One of the questions to which our examples provides few answers is sustainability. Other CFCs in other contexts have undoubtedly contributed to the emergence of a self-sustaining ecosystem. But here we still lack information on the long-term consequences. In other words, we ignore if the initial pulse can be enough to set up a sustainable ecosystem or if further impulses are needed.

There are also projects on a more international scale than truly transnational. In 2013, China and Singapore announced their willingness to share common projects on smart cities.[1] To reach this purpose, it was decided to adopt the methodology of the CFC developed in Singapore (OGDT, 2013). However, we could not find more detailed information about the process involved by an internationalization of this methodology.

We found two main insights in Singapore: (i) this city–state has a long-standing strategy of state intervention in economic development. This resulted in the creation of an instrument, the CFC with both legal and financial dimensions to foster the rise of predetermined economic sectors; and (ii) this framework and its duration are interesting for our program, as it allowed Singapore to reach beyond the stage of isolated hackathons to build a consistent strategy.

2.2. *Mexico — Retos Pùblicos*

The Mexican open data initiative has intentionally been conceived to give birth to an ecosystem (Escobar & Montiel, 2015) through new forms of PPPs (Truswell, 2016). It has been assessed by the OECD (OECD, 2016, p. 120) which praises Mexico's efforts to promote the creation of services through the constitution of PPPs within the framework of "Retos pùblicos", which means "Public challenges". These public challenges are actually part of an issue-driven method where a public agency asks the market to address a public issue through innovative services grounded on public open data. The OECD report stresses that the success of the initiative is due in part to the platform that centralizes the proposals. This platform indicates the stage of each challenge, and also has a list of key points that

[1] OG Digital Team (2013). *Smart City Collaboration between Singapore and China.* October 2013. Available at: https://goo.gl/Go54z8.

provides not only the main characteristics of the expected service, but also what already exists in the field. After the challenge, the finalists and then the winner are announced there.

In March 2015, the Mexico Economic Policy Program (MEPP, 2015) produced a "Diagnosis of Legal-Regulatory Framework for the Acquisition of Digital Services in Mexico". The report includes a description of public procurement frameworks, international best practices on similar challenge-based competitions, Mexico's legal framework, barriers on Mexican public ICT procurement, the evaluation process used during Retos Públicos pilot stage and recommendations to improve it.

One of the objectives of the program, by the kind of procurement process chosen, was to involve companies, especially SMEs, which are not accustomed to participate and submit proposals to public tenders.

This competition model allowed the creation of 15 applications from 75 prototypes and involved about 1,500 participants.

However, since the challenges are driven by different ministries, there is almost one assessment framework per challenge.

The documents presenting the challenges indicate to the potential participants the basic criteria used for the evaluation of their proposals. The relative weight of the criteria in the evaluation is also specified. Some criteria are very generic for this type of project: answers given to the basic problems, with respect to technical aspects, design, ergonomics and previous achievements. Criteria also include clarity of formulation, understanding of project issues, attempts to meet needs and trying to go beyond basic demand. They try to estimate the economic viability of the development of the applications. The evaluation also includes the use of public data open by the application and, interestingly, the production of new open data. This can be related to the concept of infomediary suggested by Pollock.

The first evaluation is ensured by the team of the National Strategy of Open Data, assisted by Codeando Mexico, an organization promoting the dissemination of civic tech. The first jury includes technicians, researchers, civil society representatives, etc. The proposals are evaluated and the top 10 selected for the next stage.

Five finalists are then selected and their projects are guaranteed to be published on the platform. Each finalist receives 10,000 US dollars if they meet the requirements of the challenge.

On the basis of this model, a generalized and more challenging program, Retos Mexico, has been designed aiming to build an open innovation ecosystem, with a platform opened to both public and private challenges. This is an interesting development of open data, although difficult to analyze since the platform is still in an embryonic stage.

The Mexican experience is inspiring as it aims purposely to build an ecosystem based on open innovation, with some challenges occurring in fields like infrastructure. This experience is also interesting as it reconciles two approaches of open data: transparency and economic growth.

2.3. *Ulyanovsk, Russia*

Riabushko (2015) presents an open data initiative in Ulyanovsk, Russia, where the open data readiness assessment of the World Bank was applied. It is a pioneering initiative and some features are very specific to the context as it projects to provide "both national independence and execution of open government principles". This region is not representative of the whole of Russia, especially since it has a relatively more developed IT sector.

This initiative also deals with the "demand side problem" of open data and intends to form an ecosystem, taking into account the main dimensions of an open data ecosystem identified in the literature: defining a legal framework, aligning with official or industry standards, or trying to engage the reusers. The promoters of the initiative identified the most important kinds of datasets: open spending, open procurement, open registers and open inspections, including food safety, housing and public utilities. Skill issues are addressed through the creation of a Centre of Competence on open government issues.

According to the assessment, this initiative is rather successful, especially thanks to a strong internal leadership, but no open data champion was identified to be a growth driver. Otherwise, this initiative remains traditional as the stimulation is still indirect.

The main outline we noticed in Ulyanovsk is related to the barriers. In spite of their will to build an ecosystem, promoters of this initiative report difficulty related to the lack of an open data champion. It is related in part to the readiness of an economic sector, but this kind of risk applies more generally to manage the issue of private sector actors' engagement.

2.4. *Finland*

Several experiences can be of interest in Finland, as there is not a centralized national strategy, as stated by OECD, but different kinds of methods put in place to drive the reuse of open data. Here, we highlight the main initiatives.

2.4.1. *Innovative Cities Program*

Tampere participated in a program funded by a European grant: Innovative Cities Program (INKA),[2] for the Smart City and Renewable Industry parts. The main focus of the program, which lasted from 2014 to 2017, was on the companies. It aimed to help local businesses to elaborate/test some solutions in a real urban environment to solve challenges. This stage should help them confront their solutions to international competition. Similar to the Singapore CFC, it can be considered as a capacity building method since the objective is to stimulate the private sector to make it more innovative by increasing its skills.

[2] INKA. The Innovative Cities Programme. https://goo.gl/f239PL.

2.4.2. *Six City Strategy*

The six cities program addresses the need to build skills of the community of public actors in order to foster the spread of an open innovation policy. It is based on the implementation of three consistent approaches:

— Open data of the cities to feed all the products/services;
— Open innovation platforms, with the aim to create new products and services through new kinds of procurement processes;
— Open participation, cocreation.

Selection includes various processes:

— Themed calls for proposals for the different focus areas of the six city strategy;
— Focused calls for proposals, such as for educational institutions;
— Negotiation procedure between partners, which can be used to carry out a continuous application process.

The strategy foresees to set up a range of common tools and services, especially in the following fields:

— To ease the cooperation between the cities, experiences and learning are stored in an open knowledge bank, also used to analyze the challenges of the projects;
— Training courses and peer learning;
— Work together to develop data catalogues with compatible metadata;
— In selected parts, standardize open programming interface in the city's system for utilization by the ecosystem;
— Harmonize the structure of the data to be as compatible and comparable as possible;
— The cities evaluate and score the projects based on the selection criteria specific to the Six City Strategy.

2.4.3. *TEKES[3] Smart Procurement Program*

TEKES is the Finnish National Funding Agency for Technology and Innovation. It coordinated a program of Smart Procurement, leading to the launch of 10 projects based on innovative procurement methodology with a budget of EUR 60 million. The aim was to speed up the introduction of innovation and to improve access for products and services developed by SMEs, in particular through a smart procurement program. This instrument is intended to orchestrate the other programs funded by this incubator by a procurement process addressing some issues faced by traditional procurement, including, for example, a weak density of relations between public buyers and private providers, a tendency to favor lowest cost proposals — an issue also addressed by the new European Union directive — and the lack of a strategy purposely designed to link what is procured and broader policy goals.

The Finnish strategy is inspiring since it uses, at the same time, the competences of an incubator with the purpose to ease the development of start-ups and SMEs that is in line with the smart city concerns. Moreover, incentives are not based on the demand of very closely defined services, but on the proposition of challenges generically defined.

2.5. *Barcelona, Spain*

Analyzing the case of Barcelona seems relevant since this city is at the forefront in the search for innovative solutions for smart cities (Juniper, 2015). The Barcelona Growth program was launched before the transposition of the 2014 European directive on public procurement. The case remains relevant since the choices of the organizers and the problems faced are similar in the new legal framework.

The program was designed as a tool to foster innovation as the country was afflicted by the economic crisis. One of the purposes was to remove the entries barriers and foster the involvement of new participants in public procurement processes.

[3] TEKES Program, SMART Procurement, https://goo.gl/dZOWpQ.

Barcelona chose to adopt a problem solving-oriented methodology (Haselmayer & Rasmussen, 2011). Six main themes were given from reducing bicycle thefts to automatic detection and alerts of damaged road surfaces.

They also address the problem of metrics to assess each challenge and faced this question: how to reconcile the need to have a substantial and common approach with the need to adapt as much as possible to the diversity of situations.

Key challenges concerned culture change in the public agencies and problem statement preparation. The point with problem statement was to design them so that they will not tend to return to the traditional solution prescription model. The consultancy company which worked with Barcelona helped write the statements so it would be understood by companies which do not have a huge tradition of participation in procurement.

The challenges were globally provided around €1,000,000: €6,000 at least and €250,000 at most per company selected for further steps.

Beside the classic evaluation criteria and those designed specifically for each challenge, they also evaluated the fact that the competitors are new in the field of public procurement, that the solutions submitted are new or derived from existing ones and that they comply with the general requirements expressed in the calls.

They tried to limit the size of the documents describing the challenges. The procurement consisted of two phases: first there was an open ideas competition, allowing the selection of five finalists. Then there was a negotiated procedure following the legal European framework, which has been reformed since then. In this kind of procurement, the specifications are written both between the government and the finalists. The stakeholders estimate that the Competitive Dialogue procedure would be more useful.

To engage users not familiar with public procurement, the consultancy company contacted 200 companies per challenge to present the themes of the challenges and the procedure intended to facilitate their access to public markets.

Among the lessons learnt, the stakeholders reminded of the importance to determine the procurement process since almost the beginning of the project. Also, the organizers identified from this experience the need to provide some training to the different stakeholders, both on the technical aspects and on the organization of the competitions.

Our interest in the Barcelona open data initiative was that this initiative was launched at the key moment of the new European Union directive on public procurement.

3. European Open Data Incubators

Open data initiatives envisioning the demand side show convergent features with incubators programs.

Recently, the European Union commission funded two open data incubators, to promote the uptake of open data reuse and value generation.

3.1. *FINODEX — Future Internet Open Data Expansion*

FINODEX is a European virtual accelerator that selected, funded and provided support services to SMEs and web entrepreneurs building their products and services making use of FIWARE technologies and reusing open data. The main objective was to launch ready-to-market ICT products and services fostering the European ICT ecosystem.

The services offered by FINODEX covered the whole spectrum of needs for the entrepreneurs. They ranged from technical support with FIWARE and open data technologies, to business mentoring to refine the business model of their product/service or a remote channel for business coaching.

Proposals were evaluated according to their technical excellence, FIWARE usage, impact (business model, expected impact on society) and implementation capacity. FINODEX has directly injected €480,000 in a total of 48 projects. By the end of the acceleration process, a total of €4.64 million will be distributed among the more than 100 beneficiaries. Following the two open calls, 493 proposals were received (respectively, 197 and 296), where 101 were selected to enter in the acceleration process.

3.2. *ODINE — Open Data Incubator for Europe*

ODINE, Open Data Incubator for Europe, is a six-month incubator for open data entrepreneurs across Europe. The program is funded with a €7.8 million grant from the EU's Horizon 2020 program. ODINE aims to establish an industry-focused ecosystem of open data start-ups and SMEs in Europe, through capacity building mechanisms using both financial investment and coaching of participants.

Proposals are evaluated according to three dimensions: (i) idea (strength or novelty of the idea, usage or creation of open data), (ii) impact (value proposition and potential scale, market opportunity and timing, triple bottom line impact), and (iii) team and budget (capacity to realize the idea, appropriateness of the budget).

Until April 2017, 57 companies were selected in seven rounds of applications.

The open data incubators, with the support of public funding, allowed the creation of a relatively decent number of services and applications. However, they remain limited to the logic of "Smart Cities 1.0", as characterized by (Barns *et al.*, 2016): dominated by small or experimental prototypes involving separate systems and infrastructures, and failing to scale or demonstrate real benefits.

These pioneering initiatives give some insights for the design and the implementation of new programs. However, some of them face a lack of documentation and of a theoretical framework, which can be introduced through an ecosystem perspective.

4. Stimulation of Open Data Ecosystems

4.1. *Definition of an open data ecosystem*

Ecosystem lens is a research stream widely used not only to get a better understanding of stakeholders and their interactions, but to think strategically and to influence this ecosystem in a given way.

We continue this approach in this contribution. Harrison, Pardo & Cook (2012) give a research agenda and, building on Nardi and O'Day, provide a broad definition of what is an ecosystem — "a system of people, practices, values, and technologies in a particular local environment". More accurately, "ecosystems are comprised of interacting, relatively tightly connected components with substantial interdependencies. Specific components will vary from ecosystem to ecosystem". Zuiderwijk, Janssen & Davis (2014) provide a state-of-the-art open data ecosystem definitions and point out the possibility to consider open data ecosystem as a combination of various kinds of ecosystems.

Open data ecosystems have been analyzed from different scales and perspectives: nationwide (Heimstädt, Saunderson & Heath, 2014), at a local scale (Dawes, Vidiasova & Parkhimovich, 2016) and according to their temporal evolution (Heimstädt, Saunderson & Heath, 2014). Ecosystems can also be analyzed through the lens of the kinds of data and services created on them. Yet it is widely admitted that different kinds of data will lead to different kinds of services, that they do not bear the same economic value potential and can thus give rise to different kinds of ecosystems.

4.2. *Functions in an open data ecosystem*

Some typologies already encompass close approaches in the field of open data (O'Reilly, 2011), a helpful summary is provided in EUDECO (2016). They are mentioned here for record because they are similar to those used in studies devoted specifically to open data ecosystems. We can point out however the European Union Data Landscape framework, which differs from other models in that the authors refuse to draw the traditional distinction between users and reusers.

Deloitte (2012) identified five archetypes: suppliers, aggregators, developers, enrichers and enablers. Interestingly, the authors' goal was not to describe the functions of an ecosystem, but rather to provide a typology of business models. To the best of our knowledge, it remains one of the most influential papers to describe the set of functions in the open data ecosystem literature. It is adopted, for example, by Ponte (2015).

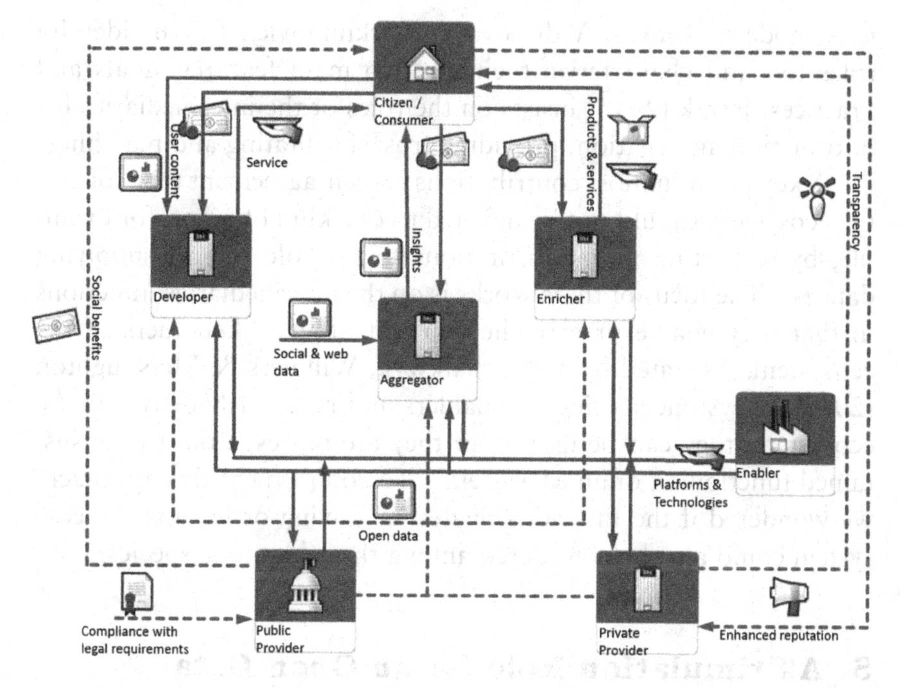

Fig. 2. Value exchanges in an open data ecosystem (Turki & Foulonneau, 2015)

Figure 2 both summarizes the state-of-the-art and our view of an open data ecosystem before its stimulation.

Lindman & Kuk (2015) propose a quite similar framework by identifying five roles in the open data value network: data publisher, data extractor and transformer, data analyzer, user experience provider and end users. For Heimstädt, Saunderson & Heath (2014), "minimal value chain within open data ecosystems consists of three elements: data suppliers, data intermediaries, and data consumers". Immonen, Palviainen & Ovaska (2014) distinguish between the roles (six main roles: data providers, service providers, application developers, application users, infrastructure and tool providers, data brokers, then refined in 22 subroles) and the services provided by the ecosystem (data provider support, data adaptation, tool support, diverse applications, contract-making, finding partners, finding services, finding information, finding markets, data validation, business models). Dawes, Vidiasova & Parkhimovich (2016) identify roles too and characterize them by two main features: goals and practices. Jetzek (2015) insists on the roles of the intermediaries for information aggregation, providing trust, facilitating and matching.

A key point in most contributions is their agreement that roles in the ecosystem should not be devoted to one kind of actor, for example, by restricting public institutions to the sole role of supplying datasets. The focus of these works is on the intermediaries' functions in that they enable or ease the work of other stakeholders of the ecosystem. As stated by Van Schalkwyk, Willmers & McNaughton (2016), "keystone species are enablers, not necessarily drivers in the ecosystem; they can be useful but they are not essential to the sustained functioning of an ecosystem". To complement this approach, we wondered if the functions related to driving or leading the ecosystem could also be considered among these keystone species.

5. A Stimulation Role for an Open Data Ecosystem

In Martin, Turki & Renault (2017), we discussed the arguments pleading for the transposition in the field of open data (O'Reilly, 2011) of the theoretical framework and of the tools developed for industry platforms. There are several differences between an open data ecosystem and an industry platform. The latter often relate success stories, like Apple iTunes or iOS and Intel processors. A platform is thus often grounded on a tangible and accurate set of products or services whereas public data and their uses represent a fuzzier set. A most basic difference is the fact that open data are offered for free to generate additional value, in consequence, the IPR regime is quite different. In spite of these differences, we showed that a transposition could be fruitful. The interest of these models consists in the richness of the real case and thus the diversity of the models, allowing different levels of integration between the components of the ecosystem, different positions and levers for the central component to influence.

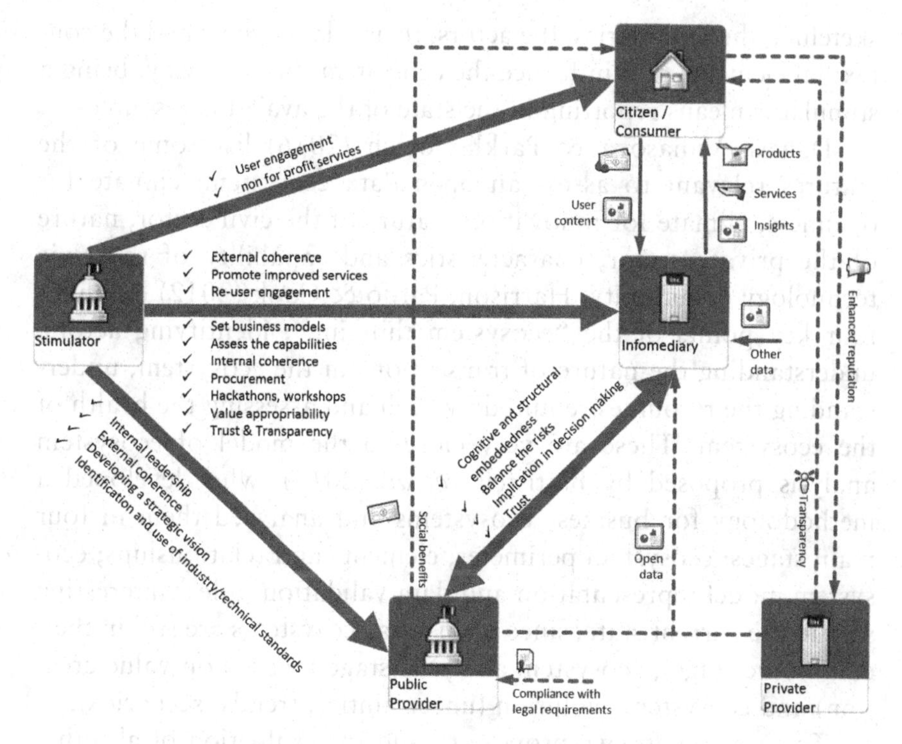

Fig. 3. Summary of stimulation function

This section lays thus the foundations of a systematic framework aiming both to describe a stimulator's role and to be a tool for planning influence on the ecosystem. Figure 3 shows the intention to balance strategy design and concrete actions. We analyze the stimulator's function in three layers: analysis, developing a strategic vision and acting to influence the ecosystem.

5.1. *Analysis of the current ecosystem and the relevant external parts*

Leadership requires knowledge of the ecosystem and most of the current works on the concept of ecosystem can be related to this stage and to the strategic analysis one. Knowing the ecosystem means sketching the boundaries, the actors, their relationships and the contextual elements that influence the ecosystem. In some way, being a stimulator means reporting in the state of the available resources.

Dawes, Vidiasova & Parkhimovich (2016) list some of the features relevant to assess an open data ecosystem: climate for openness, climate for innovation, nature of the civil sector, nature of the private sector, characteristics and capability of the civic technology community. Harrison, Pardo & Cook (2012) highlight four key points of the "ecosystem thinking": identifying actors, understanding the nature of transactions in the ecosystem, understanding the resources required by each and assessing the health of the ecosystem. These are very close to the model of ecosystem analysis proposed by Battistella *et al.* (2013), who developed a methodology for business ecosystems and analyzed them in four main stages: ecosystem perimeter, elements and relationships; ecosystem model representation and data validation (a less interesting step in the current state, since open data ecosystems are still in their embryonic stage), ecosystem analysis (stage focused on value creation) and ecosystem evolution (uncertainties, trends, scenarios).

To these points, we propose to add the evaluation of all other kinds of tangible or intangible aspects that might be included in the ecosystem or influence it. It seems, for example, relevant to include norms and values, as was done by Dawes, Vidiasova & Parkhimovich (2016). It requires having some knowledge besides the current ecosystem to identify potential stakeholders who would be attracted by the ecosystem.

5.2. *Developing a strategic vision*

This stage addresses partially strategic analysis but also strategy formulation and implementation.

5.2.1. *Definition of the ecosystem's goals*

The range of objectives can be very large: direct and indirect job creations after the creation of new kind of economics activities, capacity building and strengthening of the economic framework emergence of services that private economic activity neglects, and easing the spread of new approaches (e.g., data analytics) or of new technologies. Compared with the recommendations of the state-of-the-art, we would insist on the necessity if not adopting a prospective approach to set the ecosystem in a long-term perspective. A narrow issue is to watch the external evolutions that require an evolution of the ecosystem over time.

5.2.2. *Shape and functioning of the ecosystem*

The stimulator can influence the general architecture of the ecosystem, and its place in this architecture and define the degree of influence it shall have on the other parts.

Place of the stimulator in the ecosystem

For a given actor, this role depends on the constraints faced, its resources, its goals and its will to be applied in the network. It also depends on the internal alignment mechanisms to keep the congruence between the actor's internal organization and the whole ecosystem.

Influence of the boundaries, and shape of the ecosystem, and the nature and the strength of the relationships between the actors

As open data ecosystems are still underdeveloped, we argue that the priority concern should be the growth of this ecosystem and that these aspects should be envisioned through emergence mechanisms. This is consistent with Gawer & Cusumano (2013), who point out a limit of current research on platforms as "it takes for granted the existence of the market that transact through the platform". Thomas (2013) engaged in an analysis of ecosystem emergence through the case of digital service ecosystems. The author distinguishes three stages of emergence — initiation, momentum and control — and four main mechanisms — resource activities, technological activities, institutional activities, and contextual activities and their associated subactivities. We propose to integrate the influence on the ecosystem's goals and configuration in a broader model, which is the "orchestration process".

5.2.3. *Orchestration processes serving a strategic vision*

The orchestration process's model was developed by Nambisan & Sawhney (2011). In this section, we intend to present its components and discuss systematically the relevance of their transposition to the stimulation of an open data ecosystem. This concept requires choosing between two approaches. In an integrator model: "a hub firm defines the basic architecture for the core innovation and then invites network members to design and develop the different components that make up this core innovation". In a platform leader model, the "hub firm defines and offers the basic innovation architecture, which then becomes the platform or the foundation for other network members to build on through their own complementary innovations". Although open data seem to fit better the model of the leader, some dimensions are more adapted to the prior model. Moreover, it matters from the degree of implication aimed by the stimulator. In consequence, both are envisioned here.

The orchestration process acts on two main dimensions: innovation design and network design. Leveraging innovation design consists in acting on the assets. Applied to open data, basic elements are data, whether raw or refined. One can also include aggregated data and all intangible dimensions — as knowledge — required to analyze and reuse these data. Network design means the arrangement of the relationships of the various entities inside the ecosystem. Three main levers of an orchestration process are managing innovation leverage, managing innovation coherence and managing innovation appropriability. The authors described these processes as the interplay of more basic mechanisms, and five are discussed in Table 1.

Table 1. Main elements for innovation and network design, and issues for an open data ecosystem

Domain	Definition (Nambisan & Sawhney, 2011)	Potential application to open data
Modularity	"degree to which the network's innovation architecture has been decomposed into independent or loosely coupled modules and the interfaces that connect those modules have been specified and standardized"	Open data ecosystems being still emergent, their components are naturally loosely coupled. The stimulator has to ease the emergence of interdependencies among the members to give rise to a (certain) tight coupling. Last part on interfaces = open API?
Structural embeddedness	"how well the network members are linked (directly or indirectly) to one another, that is, it captures the overall connectedness of the network structure"	Very weak connectivity since there is a lack of interdependencies
Cognitive embeddedness	"degree of shared cognition among the network entities, that is, the extent to which members are connected to one another through shared vocabulary, common representation and interpretation schemes, and overlapping domains of knowledge"	Importance to adopt industry standards (formats, vocabularies ...) already identified concerning the release of data
Structural openness	Permeable boundaries or not	Basically high openness for the data, more complex for the other products of the ecosystem
Decisional openness	"degree to which the locus of innovation decision-making is diffused in the network"	Rather weak. Best practices advise to put in place a mechanism allowing to capture the data needs of the reusers

Managing innovation leverage

It requires envisioning both modularity and network openness to foster value creation capabilities and the attraction of new actors in the ecosystem. The means consist in "sharing or reuse of technologies, processes, intellectual property, and other innovation assets". There is a paradoxical and inverse relationship between innovation leverage and structural openness since open boundaries often give rise to an erosion of the links and conversely a need to concentrate all decision making, so there is a need of a "tiered decision-making structure whereby members that play more important roles in innovation leverage gain more say in such decisions".

If we transpose this to an open data ecosystem, some elements were already given, such as the necessity to know the ecosystem, and also the task to achieve subtle balances which will encourage the participants to share their assets. We notice equally that the paradox operates fully. At least concerning the assets that are in the form of public data already released, structural openness is the widest whereas the links between the components of the ecosystem are tenuous and even non-existent; these may help a stimulator define priority interventions. The idea that those who invest the more in their assets should be more influential in the decision-making process is interesting but if a public agency is the stimulator, it might raise some concerns about the fairness of the competition. However, the equal treatment principle is not bypassed if access to the data is equally granted for all the stakeholders. Even in the frame of public procurement, there are procedures, such as competitive dialogue, which can be related to this idea.

Managing innovation coherence

There are two scales of innovation coherence to manage:

1. Internal coherence is the "alignment of the innovation tasks, components, and interactions of the members within the network". Risks of a lack of internal coherence ("process delays, design redundancies, technological incompatibilities, higher innovation costs, and inferior performance") seem to be a good

representation of the current open data ecosystem. It is enough to look at the markets of applications reusing open data to find a significant duplication. To increase internal consistency, the authors emphasize the role of modularity and particularly of functions, such as communication and coordination, as can be workshops or hackathons.

2. External coherence, "coherence between a network's innovation goals and architecture and the external technological and market context", is less obvious in an open data context; it necessitates keeping attention on technical changes (including the standards) or market changes.

Managing innovation appropriability

A wide range of studies focused on pricing as an incentive to increase participation in the ecosystem. Open data being accessible and reusable for free, charging does not appear to be the main driver, but remains partially relevant since other assets — processes, knowledge and enriched data can still be charged and are supposed to be the basis of value creation and capture. The more generic concept of innovation appropriability appears to be better suited. Innovation appropriability refers to the "mechanisms available for partners to appropriate value from their innovative contributions". If the appropriability regime is unbalanced, the more assets a firm will have, the more it will be unwilling to participate in the ecosystem, or even if it participates, the more it will tend to show a conservative behavior. The key element is to build a "trust-based environment", through mechanisms at the interplay between structural embeddedness and decisional openness, the first one to "minimize undue appropriation of value without sacrificing the intensity of knowledge sharing". Making room for other actors in the decision-making process is also a means to increase trust in the ecosystem.

From the open data perspective, it echoes the idea that the use of open data should not be related to a requirement to share similar products or services built from these data to increase the ecosystem's attractiveness. If the stimulator envisions to go beyond the mere publication of data, for example, through the funding of infrastructures or the development of specific services, it has to find a balance in terms of intellectual property rights; it requires at least a minimal openness of the products to allow the emergence of interdependencies and the leverage of assets by each other and on the other hand the need to manage a capture of the value large enough to keep the stakeholders inside the ecosystem.

We propose to go further in the interpretation to consider a trust-related issue. Heimstädt, Saunderson & Heath (2014) present a potential risk of asymmetry between public bodies and other organizations. If public bodies choose to stop their open data initiatives, their central missions would not be directly jeopardized, whereas other organizations who would have significantly invested in the reuse of assets would see ruination of their efforts and resources. This asymmetry weakens one of the basic dimensions of an ecosystem, the existence of interdependences between the components. What is considered as an interdependency by a central organization may be considered as an undue dependency and a risk by another actor. This is the risk of one-sided dependency that the government should prevent. We think that sustainability could be considered here as a subset of trust.

Figure 3 summarizes the potential target areas and the levers available for a public body aiming to stimulate a given open data ecosystem. We plan to refine and instantiate this model following the main insights discussed in the last part of this chapter.

6. Public Actor Influencing an Open Data Ecosystem: BE-GOOD Approach

6.1. *Arguments for public bodies acting as stimulators*

Leadership to promote open data inside the government is already accepted, at least to bypass the consequences of hierarchical organizations (Harrison, Pardo & Cook, 2012). But envisioning public bodies as stimulators of the ecosystem may be seen as a paradox since they have been broadly criticized. The question is meaningful as there cannot be a leadership if the actor aiming to endorse this function is not recognized as legitimate to do so by the other stakeholders (Thomas, 2013).

However, the government is not the sole decision maker and being a stimulator does not automatically lead to an internal monopoly. Some actors like academics or non-profit organizations like ODI can assist in the steering of the ecosystem.

The government has already a central position as it defines the legal framework and is a data provider, as stated by Van Schalkwyk, Willmers & McNaughton (2016).

Managers of open data platforms face the question of economic benefits. Accurate means to assess this impact, through return on investment estimates or other kinds of indicators, are still a pending issue. There is a need for reliable and iterative instruments to measure these impacts, which are necessary to ensure the internal support of the open data initiatives. This evaluation task is consistent with the first component of the stimulation function discussed in Section 6.2 and reinforces the relevancy of a public actor to assume this role.

We have suggested above to include the emergence mechanisms identified by Thomas (2013) in our model of ecosystem stimulation. A public actor may not be relevant for each subactivity, for example, technology activities fall outside its competence, but it holds an interesting position for resource activities (an agency in charge of a data portal is likely to have the necessary legitimacy), institutional (legitimacy to establish local rules) and contextual activities (for example, regulatory activities or organization of events).

We contend that these elements provide good arguments for a public body to act like a hub firm and it in consequence is the best place to orchestrate an open data ecosystem.

6.2. *Stimulation of an open data ecosystem by a public actor*

To implement an influence strategy, our literature review allowed to identify several levers. Data publication is the first component of a stimulation strategy and it requires a strong internal leadership (Lee, 2014). The obviousness of this remark shall not hide the success of the release of a large number of datasets since the first open data initiatives.

Legal framework is fundamental too, but few public bodies have a real impact on its definition. There is a specificity in an open data context, since the rights are the lowest possible on the data. If public bodies participate in the development of products or services, it is easier for them to act on intellectual property rights and inspire themselves with the strategies established by hub firms.

Communication/dissemination, being direct or indirect, thematic workshops and apps competitions or hackathons are also helpful factors to influence the ecosystem. The evaluations of open data initiatives have shown that these actions are useful but, in most cases, were not sufficient. The model of hackathon shows some of the following limits:

- There is a lack of attention given to value capture and it is a cause of failure of civic apps competitions (Lee, Almirall & Wareham, 2016).
- In the hackathons, there are often very general themes on which the stakeholders are called to create and develop their solutions.
- One does not master the reuse made of the data, which corresponds to the original definition of the open data, but does not ease the strategic steering of the ecosystem.
- This is not the best framework to engage in long-term collaborations with the private sector, which in most cases, does not deepen or even update its services.

6.3. *BE-GOOD: Insights to implement the stimulation of an open data ecosystem*

BE-GOOD[4] is an ongoing program started in 2016 which aims at accurately defining the conceptual roots and the features of a

[4] Building an Ecosystem to Generate Opportunities in Open Data, cofunded by the European Regional Development Fund Interreg North-West Europe, http://www.nweurope.eu/begood.

stimulation function and at instantiating this function through several prototypes. BE-GOOD uses public procurement procedures as vehicles to stimulate the ecosystem. BE-GOOD is a pioneering program aiming to unlock, reuse and extract value from PSI to develop innovative data-driven services in infrastructure and environment domains. The BE-GOOD approach is based on identifying "challenges" for public sector service delivery that could be addressed through better use of data. It will then source solutions from the marketplace. This approach is new for all partners involved in the program, which mostly relied on long-term partnerships with solution providers and only had local/national links with solution providers. By a novel demand-driven approach, starting with public service delivery challenges common for public bodies across northwestern Europe to unlock appropriate datasets and engage with the marketplace, BE-GOOD shows ambitions consistent with a strategic use of procurement.

A basic assumption of this program is that an ecosystem perspective allows to identify the causes of its current weakness, the main being the low frequency of interactions between its components.

6.3.1 *Innovation and procurement*

We show in this section the new opportunities provided by the European Union directive on public procurement (2014) and the relevance to explore this framework as a vehicle for stimulating open data ecosystems and channeling stakeholders' interventions.

Public procurement in Europe is currently switching from a procedural perspective, where procurement was seen as a commodity process, to a strategic perspective where public procurement is seen as a trigger to develop public strategies, such as innovation, sustainability and social responsibility.

The European Union directive on public procurement

The last version of the European directive on public procurement (Directive 2014/24/EU) allows new forms of public procurement suited to innovation fostering among public and private partners.

Traditional public procurement procedures (open or restricted procedures) follow a quite rigid call for tenders approach where the public organization has to define beforehand all its needs and then assess the relevance of the submitted bids toward these needs. This approach is not well suited to cases where organizations are not able to define in advance their needs or imagine future services based on the opening of their data. Although exceptional procedures including negotiation are possible (competitive procedure with negotiation), such procedures are restricted to a very limited set of cases (emergency, small budgets...), which still do not fit to the development of large-scale open data-based services.

The previous version of the European Union directive on public procurement (issued in 2004) already provided an instrument to allow a form of iterative definition of needs and proposals: the competitive dialogue. This procedure allows the public procurer to iterate with several bidders in order to define both its own requirements and the bidders' proposals. In this procedure, the public organization has to define beforehand its requirements as well as the criteria it will use to assess iteratively the bidders and their proposals. Then the public organization can negotiate with bidders in order to refine their proposal. To ensure competition at least three participants shall be invited. The public organization can iteratively eliminate the suppliers during the dialogue as long as its decision is based on the criteria defined beforehand. As soon as the public organization has identified the solutions that can match its requirements, it can close the dialogue phase and ask the suppliers still in competition to submit a proposal. Eventually the public organization can negotiate the best proposal in terms of the quality–price ratio as long as the fundamental terms of the proposal are not modified. But, such a procedure did not seem to be largely used for innovation development (PricewaterhouseCoopers, 2015).

The 2014 directive went a step further by taking into account innovative approaches of public procurement used in some European member states, such as Small Business Research Initiative (SBRI) in the United Kingdom or its equivalent in the Netherlands. Thus, the 2014 directive proposes the innovation partnership instrument.

This procedure is dedicated to the acquisition of "innovative products and services that cannot be found on the market". In this procedure, the public organization defines its requirements and in particular the expected performance levels and maximum costs for the product or service, as well as the award criteria. Such award criteria shall be focused on R&D and implementation of innovative solutions. To ensure competition at least three participants shall be invited. Then in several steps with intermediary targets, the public organization progressively procures R&D activities for the development of new service/product and eventually procures the developed service when it becomes ready to be commercialized. The time frame for the innovation development and the development costs shall be proportionate to the degree of innovation sought by the public organization.

Last and specific to the case of innovation related to data processing, the (design) contest is an alternative form of procurement defined in the 2014 directive. Contests usually apply to the procurement (or the reward) of artistic works but also apply to some forms of design works (e.g., architectural and data processing). Disregarding the case where the contest outcome is a prize, the contest is a two-step procurement approach. First, a call for contest is organized, an independent jury is appointed with awarding criteria, candidates apply and may be invited to submit their proposals, the jury anonymously assesses the proposals and the jury's decision cannot be challenged. Then the public organization switches to a competitive procedure with negotiation and negotiates the supply of services with the winner(s) of the contest.

Transorganization procurement

Besides the reform of procurement processes, transnational procurement is promoted in the frame of the European Union to bypass the risks of fragmentation of demand and the lack of incentives through regional cooperation. But there are a lot of issues to address. Some European projects have considered the issue of a transnational procurement, among them PICSE, Stop & Go, Thalea or PAPIRUS projects.

Transnational procurement can be considered as a subset of joint procurement. Within a single country, joint procurement means "combining the procurement actions of two or more contracting authorities. The key defining characteristic is that there should be only one tender published on behalf of all participating authorities". In the case of a transnational procurement, another constraint is the need to comply with the requirements of the national legal systems.

The BE-GOOD program intends to encourage transnational procurement by taking into account some predictable benefits and downsides. The following are the benefits:

— Lower cost prices than in separate procedures;
— Mutualization and therefore reduction of administrative costs;
— Better skills and expertise (inside and outside the organizing authorities);
— Increased number of participants because of a larger ecosystem;
— More sustainable solutions, because of a larger market size;
— To develop transnational cooperation and methods;
— To progress in the field of transnational procurement.

And the potential downsides would be:

— Need for a careful alignment of requirements;
— Consideration of different economic and political agendas;
— Asymmetries in the transposition of the European directive and more generally different legal environments, thus leading to increased legal risks.

Previous initiatives have already explored this approach and provide several best practices. The Stop & Go program aims to deploy innovative solutions for elderly people through a transnational procurement. The goal of the program is also to propose a European Specification Template for this kind of procurement, but we still lack information about practical implementation and lessons learned.

During the PAPIRUS program, since there were projects in different European countries, all almost at the same degree of maturity, various approaches were envisaged to build a transnational approach, trying to mutualize as many elements as possible while respecting the local constraints. Following these principles, three approaches have been studied:

— "Forming a new procuring body consisting of all four pilots;
— All four pilots acting as a buyers group;
— Adapting a common frame for the technical and organizational implementation of the different procurements in all pilots and coordinating parallel tenders".

The PAPIRUS project's members finally thought that the only viable solution was to launch separate procurements for each pilot project, while respecting a common framework and coordinating projects. This example shows that even in the cases of common maturity projects, it is still not easy to set up a full kind of joint procurement.

However, more flexible and intermediate forms make it possible to benefit from a collaboration. Depending on the objectives of each challenge owner, it is necessary to define a common (and minimal) ground and to set up the actions to be taken jointly and steer of cooperation between entities.

6.3.2. *BE-GOOD approach of ecosystem stimulation*

Having taken into account the opportunities of the regulatory framework for public procurement, we show how it may be a lever to stimulate the ecosystem according to our stimulation model. BE-GOOD will explore each dimension and we focus here on the main insights.

In this approach, a public agency tries to engage with the reusers on long-term collaboration, the opposite to most of hackathons. It can choose to attract reuser's attention and creativity on a given set of data, whether it considers that they are neglected or that their use could foster value creation. It may choose to engage with the reusers in a problem-solving approach, where the public body defines the problem and asks the market to suggest innovative solutions. In both cases, civil servants will hardly have from the beginning a clear idea of what datasets or what problem they want to highlight to the reuser's community.

When a public body matures an idea intended to be submitted as a problem to the reusers of open data, the civil servants will have to put themselves in the place of the potential reusers and experience of the problems they will face, e.g., data discoverability, data quality or other intellectual property concerns. A public body cannot be considered as a whole, the siloed data problem is encountered by the general public as well as by civil servants. Thus, public procurement can be a way to think as a reuser might do, and so to develop a better knowledge of the barriers and risks they will face. At the same time, engaging in public procurement often requires a phase of market research (Immonen, Palviainen & Ovaska, 2013) to identify existing solutions on the market and thus potential data reuser to engage with.

All these elements are consistent with the first part of the stimulation framework exposed in the second part, knowing the ecosystem.

Likewise, we have shown both the requirement to think about current open data ecosystems in terms of emergence and the opportunity for a public body to carry out the related activities. Public procurement seems well suited to leverage the orchestration processes defined in Section 5.2.3 of our framework and derived from Nambisan & Sawhney (2011). Although being constrained by a legal framework paying considerable attention to the fairness of the competition, existing procedures are consistent with the engagement of a wide range of actors and thus with the network design of the ecosystem. A public body is well positioned to set new interdependencies since its goals are related to the common good and the growth of the ecosystem, being agnostic concerning the chosen technologies and with a legal framework preventing the privilege of one over other companies. Moreover, a public body is well positioned to turn the network to a transnational dimension. This is a paradox, as a public body often has geographically limited attributions, but it can succeed through some kinds of instruments allowing a transnational collaboration, and through the generation of a transnational demand, which can be helpful to give some ideas for the critical mass to be profitable.

Long-term user engagement combined with an issue-driven procurement may be a good way to increase the degree of modularity between the components of the current ecosystem. Public procurement seems especially suited to manage innovation appropriability. Public intervention in the creation of service can strengthen trust in the ecosystem. Above all, this role allows the stimulator to balance intellectual property rights in order to enable both appropriation of value and the sharing of intellectual property, thus leading to the emergence of interdependencies and finally increasing strength of links inside the ecosystem.

We argue that a public procurement approach is suitable to bring a stimulator closer to the operating mode of a hub firm and is a means to ease the emergence of the ecosystem.

7. Conclusion

This chapter presented both empirical and theoretical perspectives on the stimulation of an ecosystem. An overview of recent and current initiatives to promote second generation stimulation of open data reuse through the emergence of sustainable ecosystems allowed to summarize lessons learned and related best practices.

Then, we suggested a theoretical framework combining these elements in a unified system. As a conscious actor of the existence of the ecosystem, the stimulator can set goals for the ecosystem, not only on what it can produce, but also on its form and functioning. Through orchestration processes, an actor endorsing this function can manage risks in the ecosystem, not only the risks carried by the network, but also to mitigate and prevent the risks perceived by other actors. This role should foster current ecosystems, which are still fragile, incomplete and suffer from a lack of integration. This framework could help identify and develop the interdependencies, to increase the density and the strength of the relationships among the components of the ecosystem.

This model will be refined and instantiated within BE-GOOD, an ongoing initiative aiming to test this approach using public procurement as a vehicle. This program necessitates formalizing a role encompassing a stimulation function allowing public bodies to influence the ecosystem through various tools, both organizational and financial and specifically through the means of public procurement. Although public procurement is strongly tied by a legal framework attentive to the rules of equal treatment, access to information and all matters relative to fair competition, there is still room to use public procurement in a strategic manner.

Further research will encompass specific influence actions and address points like the skills required by civil servants to be the stimulators, what kind of public procurement procedure would be the most suitable to draw value from open data, and what are the evaluation criteria able to foster the creation of value; it will further elucidate how to avoid the risks associated with these kinds of influence mechanisms.

References

Barns, S., McNeill, D., Cosgrave, E. & Acuto, M. (2017). Digital infrastructures and urban governance. *Urban Policy and Research*, *35*(1), 20–31, Special Edition on Critical Infrastructure.

Barry, E. & Bannister, F. (2014). Barriers to open data release: A view from the top. *Information Polity*, *19*, 129–152.

Battistella, C., Colucci, K., De Toni, A. F. & Nonino, F. (2013). Methodology of business ecosystems network analysis: A case study in Telecom Italia Future Centre. *Technological Forecasting and Social Change*, *80*(6), 1194–1210.

Chan, C. M. L. (2013). From open data to open innovation strategies: Creating e-services using open government data. In *46th Hawaii International Conference on System Sciences*, Wailea, Maui, HI, January 2013, pp. 1890–1899.

Dawes, S. S., Vidiasova, L. & Parkhimovich, O. (2016). Planning and designing open government data programs: An ecosystem approach. *Government Information Quarterly*, *33*, 15–27.

Deloitte (2012). *Open Data Driving Growth, Ingenuity and Innovation.* Deloitte analytics paper. Available at: https://goo.gl/XrQR4S.

EUDECO (2016). *Report on the Socio-Economic Analysis*. Modelling the European data economy. Available at: https://goo.gl/nvgbdX.

Foulonneau, M., Martin, S. & Turki, S. (2014a). How open data are turned into services? In *International Conference on Exploring Services Science*, Geneva, Switzerland, February 2014, pp. 31–39. Springer 2014 Lecture Notes in Business Information Processing.

Foulonneau, M., Turki, S., Vidou, G. & Martin, S. (2014b). From open data to data-driven services. In *14th European Conference on eGovernment*, Brasov, Romania, June 2014.

Gawer, A. & Cusumano, M. A. (2014). Industry platforms and ecosystem innovation. *Journal of Product Innovation Management*, *31*(3), 417–433.

Harrison, T. M., Pardo, T. A. & Cook, M. (2012). Creating open government ecosystems: A research and development agenda. *Future Internet*, *4*, 900–928.

Haselmayer, S. & Rasmussen, J. (2011). *Navigate Change: How New Approaches to Public Procurement will Generate New Markets*. Ministry of Enterprise and Labour Catalonia Competitiveness Agency.

Heimstädt, M., Saunderson, F. & Heath, T. (2014). From toddler to teen: Growth of an open data ecosystem. *eJournal of eDemocracy and Open Government*, *6*, 123–135.

IMDA (2016). *Opening Up New Smart City Opportunities for ICT Firms*. Press release, Singapore, November 2016. Available at: https://goo.gl/I5kdxb.

Immonen, A., Palviainen, M. & Ovaska, E. (2014). Requirements of an open data based business ecosystem. *IEEE Access*, *2*, 88–103.

Jetzek, T. (2015). *The Sustainable Value of Open Government Data: Uncovering the Generative Mechanisms of Open Data through a Mixed Methods Approach*. Frederiksberg: Copenhagen Business School.

Juniper Research, Smart Cities: On the Faster Track to Success. February 2015, Available at: http://goo.gl/SmQq8W.

Land Transport Authority of Singapore (2015). *Call for Collaboration — Multi-Modal Journey Planner for Singapore*. October 2015. Available at: https://goo.gl/ePmBhr.

Lee, D. (2014). Building an open data ecosystem: an Irish experience. In *Proceedings of the 8th International Conference on Theory and Practice of Electronic Governance*, Guimaraes, Portugal, October 2014, pp. 351–360.

Lee, M., Almirall, E. & Wareham, J. (2016). Open data and civic apps: First-generation failures, second-generation improvements. *Communications of the ACM*, *59*(1), 82–89.

Lindman, J. & Kuk, G. (2015). From open access to open data markets: Increasing the subtractability of open data. In *48th Hawaii International Conference on System Sciences*, Kauai, HI, January 2015, pp. 1306–1313.

Magalhães, G. & Roseira, C. (2016). Exploring the barriers in the commercial use of open government data. In *9th International Conference on Theory and Practice of Electronic Governance*, Montevideo, Uruguay, March 2016, pp. 211–214.

Martin, S., Foulonneau, M., Turki, S. & Ihadjadene, M. (2013). Open data: barriers, risks and opportunities. In *13th European Conference on eGovernment*, Como, Italy, 13–14 June 2013, pp. 301–309.

Martin, S., Turki, S. & Renault, S. (2017). Open data ecosystems: introducing the stimulator function. In *International Conference on Electronic Government and the Information Systems Perspective*, Lyon, France, August 2017, pp. 49–63.

Mexico economic policy program, third annual report, July-Sept 2015. Available at: https://goo.gl/Rtv6YZ.

Nambisan, S. & Sawhney, M. (2011). Orchestration processes in network-centric innovation: Evidence from the field. *The Academy of Management Perspectives*, *25*(3), 40–57.

OECD (2016). *Open Government Data Review of Mexico: Data Reuse for Public Sector Impact and Innovation*. OECD Digital Government Studies, June 2016. Available at: https://goo.gl/Pqf1dC.

O'Reilly, T. (2011). Government as a platform. *Innovations*, 6(1), 13–40.

Ponte, D. (2015). Enabling an open data ecosystem. *ECIS 2015 Research-in-Progress Papers*, 55.

PricewaterhouseCoopers (2015). *Study on "Strategic Use of Public Procurement in Promoting Green, Social and Innovation Policies"*. Report for the European Commission, p. 134.

Riabushko, A. (2015). Open data initiative to challenge the demand side problem. In *Proceedings of the 2015 2nd International Conference on Electronic Governance and Open Society: Challenges in Eurasia*, St. Petersburg, Russian Federation, 24–25 November 2015, pp. 10–16.

Thomas, L. (2013). Ecosystem emergence: An investigation of the emergence processes of six digital service ecosystems. PhD Thesis, Imperial College Business School.

Truswell, E. (2016). Open data, innovation and public–private partnerships. In *Exploiting Open Data Conference*, Brussels, Belgium, October 2016, Available at: https://goo.gl/6WTQXo

Turki, S., Martin, S. & Renault, S. (2017). How open data ecosystems are stimulated? In *Proceedings of the International Conference on Electronic Governance and Open Society: Challenges in Eurasia*, St. Petersburg, Russia, 4–6 September 2017, pp. 179–187.

Turki, S. & Foulonneau, M. (2015). Valorisation des données ouvertes: Acteurs, enjeux et modèles d'affaire. In *5th Conference on DocSoc*, Rabat, Morocco, 4–5 May 2015, pp. 113–125.

Van Schalkwyk, F., Willmers, M. & McNaughton, M. (2016). Viscous open data: The roles of intermediaries in an open data ecosystem. *Information Technology for Development*, 22, 68–83.

Zuiderwijk, A., Janssen, M. & Davis, C. (2014). Innovation with open data: Essential elements of open data ecosystems. *Information Polity*, *19*, 17–33.

Chapter 3

The Ecosystem of Open Data Stakeholders in Sweden

Serdar Temiz, Marcel Bogers[†] and Terrence E. Brown**

*Department of Industrial Economics and Management,
KTH Royal Institute of Technology, Stockholm, Sweden
[†]Department of Food and Resource Economics,
University of Copenhagen, Denmark
Haas School of Business, University of California, Berkeley

Open data is trending among both practitioners and researchers. Public authorities are opening their datasets but "open data is a means not an end, and releasing the data is a small step in a long walk". (Broster, Misuraca & Bacigalupo, 2011, p. 62, 63). At the same time, open data does not merely cover technical subjects but also relates to policy, legal, economic, financial, organizational and cultural challenges (Attard *et al.*, 2015, p. 416). The lack of consumers using existing open data portals further indicates that more research is required to understand factors that influence participation in and usage of open data. It also sets a challenging requirement for engaging stakeholders to participate and collaborate in order to innovate the way they interact with the government and develop applications (Attard *et al.*, 2015).

Due to the varied roles of different stakeholders, it is vital to understand an open data ecosystem in order to pinpoint problems and suggest priorities for which challenges to overcome and how to overcome them. Chesbrough (2003) defines modes of the innovation process as funding, generating or commercializing innovation. Similarly, West & Bogers (2014) present a process model of obtaining, integrating and commercializing external innovations, to which they add non-recursive paths which involve reciprocal interactions with cocreation partners. We use this model as a framework to analyze the main stakeholders utilizing open data and how they interact with each other.

In our taxonomy, for the sake of our research, Open Innovation Ecosystem framework based on both West & Bogers (2011, 2014) and Chesbrough (2003) is proposed to be used and we will look at the open data ecosystem from the framework as our lens of study: a four-base process model of open innovation and modes of innovation. Sweden is used as the unit of study. It can be said that the framework is used from a Swedish perspective.

1. Open Government, Open Data and Open Government Data

The idea of open government, coined during peacetime in America after the information restrictions of the Second World War, was aimed at reintroducing transparency and accountability to the government (Yu & Robinson, 2012). One can conclude that a government can be an open government without involving open data, in the sense of being transparent. Enhanced technologies, such as Web 2.0 technologies and social media, push for new models with the intent of "bringing technology, people and government to collaborate and participate in the way government works, creates, and promotes and enacts solutions" (Bertot *et al.*, 2010, p. 2). Berners-Lee's (2010) star rating for linked open data highlights the importance of the legal and technical openness of data.

Governments are the biggest information and data collectors in most national ecosystems and also the largest providers. It is not, therefore, a surprise that discussions and movements focused on open data started with government datasets but open data is not

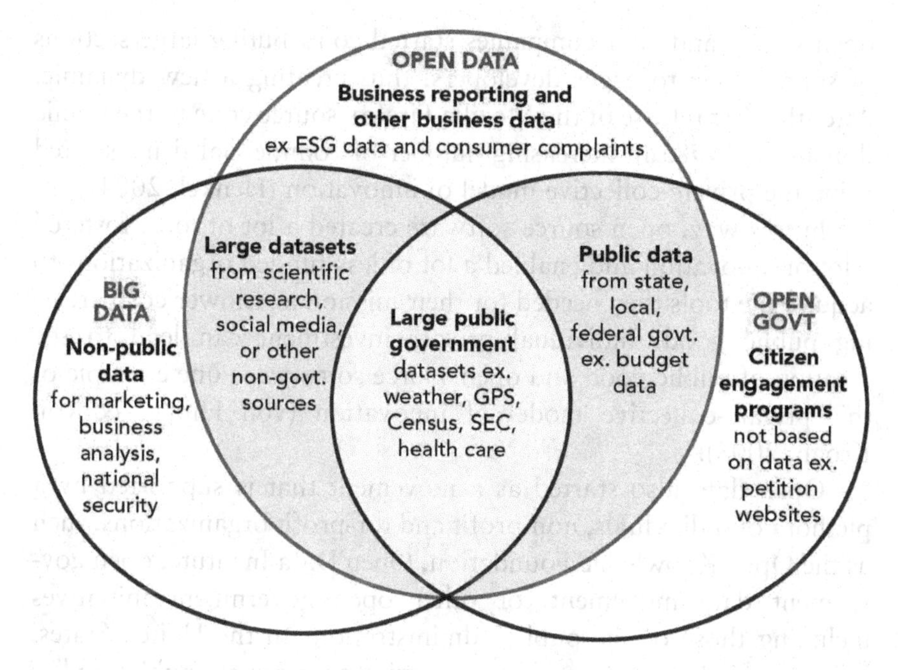

Fig. 1. Types of open data (Gurin, 2014)

exclusively open government data (OGD). As Sandoval-Almazan & Gil-Garcia (2015) also point out, the open government movement has promoted the availability of big data not only from the public sector, but also from the private and civil sectors. The term open data, however, has a generic meaning referring to the basic form and accessibility of data (Yu & Robinson, 2012). OGD refers to content that originates from public administration.

Sandoval-Almazan & Gil-Garcia (2015) state that, even though they are related, e-government, open government, open data and big data should be treated as different topics. Gurin (2014) illustrates types of open data (Fig. 1) with regard to the intersection of open government and big data.

It's interesting to look at the open source software movement as an analogy to the open data movement. Similar to open data, open source software started as a movement of technologists, hackers and hobbyists outside corporate boundaries striving for "freedom of information" (Stallmann, 1992). At first, the data was created as open source and later companies started contributing large sections of source code to eager developers, thus creating a new dynamic. After the first release of the Mozilla Firefox source code to the public domain in 1998, an increasing number of commercial firms started using the private-collective model of innovation (Henkel, 2004).

In this way, open source software created a lot of jobs, fostered a lot of innovation and enabled a lot of less-funded organizations to acquire the tools they needed for their mission at a lower cost, creating public good. Individual private investment can lead to the creation of public good and open source software is one example of this private-collective model of innovation (von Hippel & von Krogh, 2003).

Open data also started as a movement that is supported by a plethora of individuals, non-profit and for-profit organizations, such as the Open Knowledge Foundation, Open Data Institute, open government data movement, or other open government initiatives including those of the public administrations in the United States, UK and Sweden. Governments are trying to motivate public bodies to open datasets as well as private parties to harness these datasets to create new usage.

Another perspective on open data is to view it as infrastructure, similar to how you would look at roads or the energy grid, both as an enabler and as content (Poikola, Kola & Hintikka, 2010). For example, in Sweden, the electricity service was, initially, provided by the state and currently citizens can buy electricity service in a free market from different private companies.

1.2. *Open data stakeholders*

Freeman (2004) defines a stakeholder as "any group or individual who is affected by or can affect the achievement of an organization's objectives". The primary goal of the stakeholder theory was to assist managers to identify and manage stakeholders. Even though there are "several obvious weaknesses of the book", stakeholder theory has become the underpinning of management literature across different disciplines (Freeman, 2004).

It is further important to note that the roles of stakeholders in an open data ecosystem are not static and may change over time. Conditionally, and depending on the specific actions of other stakeholders, the roles of people or groups can change or in some cases, stakeholder roles may become blended.

1.3. *Open data ecosystem*

To grasp the origin and usage of this supply chain of open data, and to perform a solid analysis, an appropriate way of looking at the complex web of actors and consumers needs to be fostered. The complexity and patterns are not dissimilar from those of a biological, organic ecosystem, and as this work suggested previously, it is suitable to use methods from the natural sciences when examining economic, social and socioeconomic phenomena.

Therefore, considering the prior scholarly work, it is vital to understand the ecosystem of open data at various stages from local communities all the way to a European and global scale, and to document what is currently happening in this interdependent open data ecosystem, and to identify its key stimuli, chart its natural dynamic, and map its complex interdependencies, so as to identify challenges and obstacles and be better able to develop approaches to surmount these.

Ohemeng & Ofosu-Adarkwa (2015) divide open data into two groups on the demand side (citizen, user) and supply side (public bodies), stating that most of the research is done on the supply side of data.

Robinson *et al.* (2009) suggest that governments should not develop or select best practices, but instead should rely on private parties in a vibrant marketplace of idea engineering to discover what works best, albeit this research does not provide a suggested method for achieving this.

Immonen, Palviainen & Ovaska (2014) conducted interviews with Finnish companies as a basis for their research, and although their results cannot be generalized, they present significant comparative value to similar research in other countries. Traditional models based on a simple value chain are not sufficient to explain new digital economics, where there is no straightforward value chain, and "where services, data and providers are mixed in the picture" (Bonina, 2013, p. 12). Scholars state that interests in open data and open data ecosystems are high but there is an observed lack of usage of open data portals, therefore further research work is required to understand participation patterns and the relationship between stakeholders of the open data ecosystem (Attard *et al.*, 2015; Immonen, Palviainen & Ovaska, 2014). Attard *et al.* (2015) also state that if the target consumers of the data do not actually use it, then the objectives of the said open government initiative have fallen short.

Open data, in this sense, has clear analogy with the history of e-government, and too few studies have been conducted on both of these topics, with a particularly notable "lack of refereed, rigorous, and independent academic studies beyond a government and consultant 'grey' literature of mixed quality" (Gauld, Goldfinch & Horsburgh, 2010, p. 177; Ohemeng & Ofosu-Adarkwa, 2015, p. 420). Open data need to be evaluated from the viewpoint of citizens (Nam, 2012).

1.4. *Open innovation*

Early instances of identified open innovation comprised series of bilateral collaborative endeavors between two organizations to open up or freshen up their internal innovative processes. Today, open innovation is understood as managing multi-actor innovation communities across multiple, complex roles in the innovation process in order to attain the effective use of internal and external knowledge in every diverse entity involved (Chesbrough, 2012).

There are two types of open innovations: (i) inside-out, where idle resources are required and ideas are generated to be shared with external organizations (competitors, partners, suppliers, users, universities) for their business and business model; and (ii) outside-in where the firm or organization must first remove company barriers to external ideas and resources, and to absorb innovation and knowledge outside the contained, controlled environment within its auspices (Chesbrough & Crowther, 2006; Chesbrough, 2012). The former model has been explored less and understood less, while the latter has been explored by many scholars and firms (Dahlander & Gann, 2010; Chesbrough, 2012). Both types of innovations can have pecuniary and non-pecuniary incentives (Dahlander & Gann, 2010).

We argue that a similar framework of open innovation can be used to study open data and ultimately open government, as these coexist within the same sphere, and broadly enough within the same ecosystem.

Open innovation is a model where developers, users, influencers and basically all ecosystem actors are an essential part of the ideation and development process of services and products. In the same way, opening data can be seen as an interactive, dynamic and multilateral process of push and pull, mainly because often best ways to use the data are present outside the organization (Poikola, Kola & Hintikka, 2010). In brief, this work argues that the perceived failures of user activity around instances of open data, and cases of lack of innovation, are a result of an oversimplified value chain, and an insufficient scale of reach in its collaborative ecosystem.

2. Method of Analysis

Government websites, open data portals, online databases, press and periodical articles, official press releases and statements, official speeches, and official documents are our resources, and all these contents are analyzed relatively to open innovation, open data and e-government literature via the set framework.

2.1. *Open data ecosystem in relation to data*

If we try to define the boundaries, an open data ecosystem from a data perspective, and check how Poikola, Kola & Hintikka (2010) described stakeholders, we can filter it into three categories: open data providers, open data users and open data interpreters.

Open data providers can be either a public organization that releases public sector information or private organizations that open up their API or provide web services in order to grant third-party developers and applications access.

Open data users are organizations that are using open data to develop applications. These applications can be desktop, mobile or virtual world applications.

Interpreters are the final recipients of data to interpret and provide for their further harnessing. This form of categorization does not say much about the innovation processes themselves, but presents the ecosystem players alongside the type of data they interact with.

The following list can be divided into further sublists but for the sake of simplicity, we will avoid it.

1. Public organizations that release public sector information.
2. Private organizations that have public API or web services. Here it is important to note that a private organization can either be from an old industry (e.g., car industry) or from the IT industry.
3. Private organizations that use open data to develop 2D, 3D and mobile applications.
4. Public organizations that use open data to develop new web and mobile applications.

We have depicted all these simplified categories in Fig. 2.

2.2. *Different modes of innovation using open data*

As discussed in Section 1.3, inside-out and outside-in open innovation are two modes of transaction which provide a prism which filters interactions within a studied ecosystem. Bilateral processes are still valid, but with exponentially available data and gradually growing culture of multilateral collaboration in innovation processes, they are rarely truly bilateral.

Chesbrough categorizes the activity of companies into three areas based on their posture in relation to the innovation process: funding innovation, generating innovation or commercializing

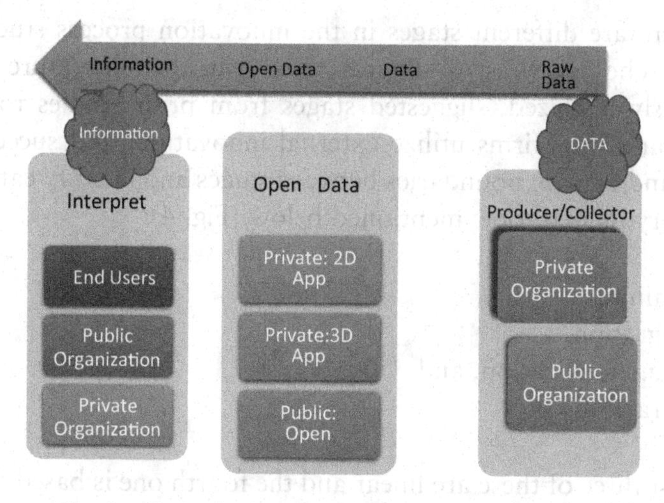

Fig. 2. Open data ecosystem from a data perspective

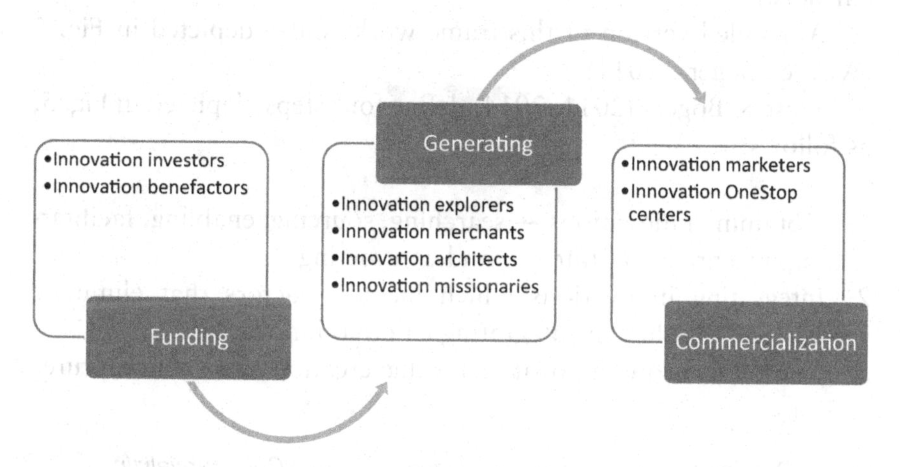

Fig. 3. Chesbrough's (2003) model of open innovation

innovation, and then lists type if organizations for each category as depicted in Fig. 3 (Chesbrough, 2003). The Chesbrough model of open innovation is a linear model.

Many firms and organizations have selected one mode of innovation to focus on: funding, generating or commercialization (West & Bogers, 2011).

There are different stages in the innovation process studied by diverse scholars. West & Bogers (2014), in their literature review previously analyzed, suggested stages from prior studies to better understand how firms utilize external innovation, and successively define and explain boundaries between stages and identify categories and every related topic mentioned below (Fig. 4):

1. obtaining,
2. integration,
3. commercialization, and
4. interaction.

The first three of these are linear and the fourth one is based on non-recursive paths that involve reciprocal interactions with cocreation partners.

A detailed version of this frame work is also depicted in Fig. 5 (West & Bogers, 2011).

West & Bogers (2011, 2014) define four steps, depicted in Fig. 5, as follows:

1. Obtaining innovations — searching, sourcing, enabling, facilitating, incentivizing, filtering and contracting.
2. Integrating innovations which includes factors that eliminate barriers that hinder integrating external innovations.
3. Commercializing innovations, value creation and value capture.

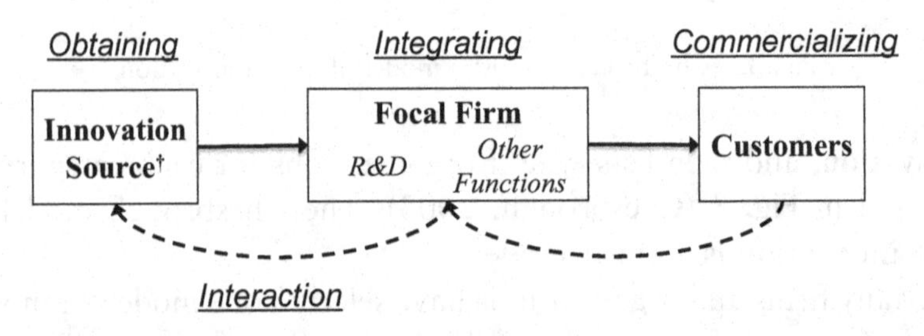

Fig. 4. Four phases of the process model for leveraging external innovation

Note: †Sources may include suppliers, rivals, complementors and customers.

Source: West & Bogers, 2014.

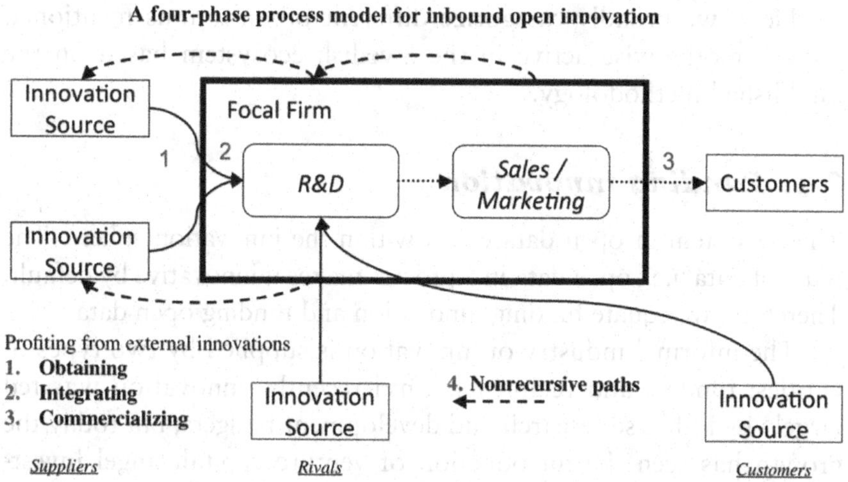

Fig. 5. A four-phase process model for inbound open innovation (Source: West & Bogers, 2011)

4. Interaction mechanism is the non-recursive path which includes feedback loops, reciprocal interactions with cocreation partners, and integration with external innovation networks and communities.

In our taxonomy, we will combine both West & Bogers (2011, 2014) and Chesbrough (2003) for the sake of our research and we will look at the open data ecosystem from these two frameworks as our lens of study: a four-base process model of open innovation and modes of innovation.

3. The Innovation and Open Data Prism in Sweden

When looking at diverse actors in Sweden through an ecosystem perspective, the unrefined relationship to open data becomes more visible. As mentioned in the Section 2 actors fully or clearly devoted to open data in Sweden offer an incomplete picture. A true taxonomy therefore needs a more indirect approach, revealing more influential instances.

Here, we broadly categorize different organizations mentioned above or otherwise active in the Swedish ecosystem based on the established methodology.

3.1. *Funding innovation*

The ecosystem of open data exists within the innovation sphere. The reuse of data, i.e., open data in a process makes it innovative by default. Therefore, we equate funding innovation and funding open data.

The informal industry of innovation is supplied by two types of entities: funders and recipients. Until recently, innovation was fed largely by in-house research and development budgets, but today the growth has seen the introduction of venture capital, angel investment, corporate investment, private equity investment and small business investment companies. These innovation funders provide alternatives to complement or supplement a research budget. The main difference according to Chesbrough is that investors are more long-term-oriented, while benefactors put emphasis on the initial steps in a research chain's output (Chesbrough, 2003, p. 38).

Companies setting themselves up as benefactors are becoming more and more frequent. They leverage money on high-potential nascent research efforts, and thus get the early pick of ideas which can feed the innovation processes within their sector or another sector of interest. The new dynamic in this stage which has recently emerged is the substantial rise in private foundations, especially ones backed by affluent private individuals engaging in philanthropic forays into research. Chesbrough gives the example of Larry Ellison, Chairman and CEO of Oracle, the parent company of Java, set up a foundation feeding 50 million USD annually in cancer research (Chesbrough, 2003).

While this new dynamic has been captured in recent academic works, the dynamic within specific contexts, the conditional variables, and the panorama of outcomes and consequences have not received extensive attention (West & Bogers, 2014).

In the successive sections, an attempt is made at using these two prisms to place a number of key actors within the open data and open innovation spheres in Sweden into their active open data ecosystem roles. In many cases, open data is not an exclusive activity, but these organizations and entities still play an active role within this scope.

3.1.1. *Innovation investors*

Although functioning in a free market, the Swedish State is very much of an investor in innovation, both directly and through foundations and agencies. The funding and public–private partnerships are extensive, and for this reason, it would require a further, separate study to quantify them within the realm of open innovation or open data alone (Table 1).

The Nordic innovation accelerator is an open innovation platform which invests in diverse fields, much as a venture investor. Within the Swedish ecosystem there are examples of similar open innovation platforms.

KK-stiftelsen is another good example of a Swedish innovation investor, a more “traditional” foundation that works with open innovation methods.

These three actors qualify as innovation investors within open data and e-government in Sweden. They are investing in projects in the longer term, and make plans and commitments past the initial phases of support they provide.

3.1.2. *Innovation benefactors*

Vinnova, SIDA and ISS are three diverse kinds of benefactors which have promoted start-up projects in open data and open government (Table 2). They qualify as benefactors because they serve more as incubators for the start-up phase and prefer to extend their reach rather than walk with projects in the long term.

3.1.3. *Public administration and civil servants*

The individual bureaucratic initiative seems to be a key driving force, and there seems to be a bottom-up model in public administration with good initiatives often coming from the bureaucratic initiative as

Table 1. Innovation investors.

Organization name	URL	Open data exclusive	Years active	List category
The Swedish Government	http://www.government.se/contentassets/cbc9485d5a344672963225858118273b/the-swedish-innovation-strategy	No	N/A	Government
Nordic Innovation Accelerator	https://www.nordicinnovation-accelerator.com/fi/blog/nordic-innovation-accelerator-joins-forces-with-the-key-actors-in-cleantech-in-finland-sweden-and-denmark	No	Since 2006	Business
KK-stiftelsen	http://www.kks.se/SitePages/Startsida.aspx	No	Since 1994	Non-profit

Table 2. Innovation benefactors.

Organization name	URL	Open data exclusive	Years active	List category
Vinnova	http://www.vinnova.se/	Yes, department	Since 2001	Public administration
SIDA	http://www.sida.se/English/press/current-topics-archive/20121/Open-data-means-transparent-aid-/	No	Since 1995	Public administration
ISS	www.iis.se	No	Since 1997	NGO foundation

opposed to political prerogative. While in Sweden the actual practice of opening data varies broadly between public institutions, there is a national agency which has an entire department devoted to open data innovation, Vinnova.

Founded in 2001, Vinnova is the Swedish innovation agency. As a relatively modern institution, Vinnova seeks to fund selectively in order to nurture renewal within Swedish industry and the economy.

Table 3. Vinnova funding amount (2012–2015).[1]

Years	Funded projects	Organizations receiving funding (Total)	Number of non-governmental agencies funded	Total amount (in SEK)
2012	37	33	22	14,600,780
2013	20	18	10	7,898,063
2014	18	15	9	5,567,788
2015	44	29	22	7,604,335

The agency has a special funding line for open innovation: "open data creates opportunities for innovation. Open data is relevant towards making available the best skills and the best knowledge not available within an organization in order to foster innovation. It is a way to open up innovation processes in order to make innovation possible" (Vinnova.se).

The open data section of Vinnova has three staff members, and they seek to not only fund (Table 3), but also support discussion and conferences around open data. As mentioned previously, their main efforts are to prioritize increasing public awareness around open data opportunities.

Vinnova is alimenting a substantial part of open data innovation within Sweden. It has funded nearly 100 diverse organizations working with open data and is helping diversify the open data ecosystem by focusing its support on a broader range of actors: "A necessary first step in this form of innovation is the opening of data sources. Therefore, Vinnova has a call providing support for those who want to open up their data. The next step is support to creative solutions created by interested citizens, researchers, hackers, businesses and nonprofit organizations using the data".[2]

---

[1] Collected and adapted from http://www.vinnova.se/sv/Var-verksamhet/Strategiskt-viktiga-kunskapsomraden/Tjanster-och-IKT/Oppen-innovation/Oppna-data/.

[2] http://www.vinnova.se/sv/Var-verksamhet/Strategiskt-viktiga-kunskapsomraden/Tjanster-och-IKT/Oppen-innovation/Oppna-data/.

3.1.4. *Other government agencies*

According to Sweden's OGP Action Plan 2014–2016:

"Sweden is a mature eGovernment nation with a high degree of transparency and efficiency. Sixty per cent of Swedish citizens use eServices. There are over 3,800 eServices in place (over 1000 of which are machine-to-man) and 40% of government agencies work actively with open data".

In practice, although there are a lot of good services available, and generally many important services and datasets are well engineered according to user reviews, a good percentage of this 40% is rather passive and lacks built-in usability.

The Swedish Innovation Strategy for 2020 mentions "open data" only a single time in 105 pages, and sticks to the vague notion of openness more frequently. Comparatively, it mentions the word "innovation" 423 times and "open innovation" five times. This could be taken as an illustration of how slowly open data and innovation are being linked at the rhetorical level.

A leading agency is the Swedish Environmental Protection Agency. The agency seems to have realized, very much like their colleagues in the UK and a number of other European countries, where open data is extremely pertinent to both direct innovation and business model innovation in terms of tackling environmental issues.

Björn Risinger, the Director General of the Environmental Protection Agency recently said, "Open data is everyman's right (allemansrät)" (Näsfors, 2016).

An important actor which can spread open data notions is the Swedish Association of Regions and Municipalities or SKL. They have an active open data section and actively promote the idea with their members.[3]

Lantmäteriet is the authority that maps Sweden that has tasks that include registering and securing the ownership of all properties as well as managing their borders. Lantmäteriet will change

---

[3] http://skl.se/naringslivarbetedigitalisering/digitalisering/digitaldelaktighetoppenhet/oppnadata.oppnadata.html.

license terms from CCBY to CC0, which means that the source no longer needs to be specified[4] after September 2017.

3.1.5. *Politicians*

Domestically, open data is not very much part of the current political agenda on any level in order to be considered a driving force.

The Swedish Association of Local Authorities and Regions is one of the strongest bottom-up drivers of open innovation in Sweden.

"There is a simple and fast way to get started publishing open data" (SKL Swedish Association of Local Authorities and Regions, 2015). "We hope that we can encourage municipalities to increase publishing information from their own operation systems by extension, said Peter Mankenskiöld, coordinator manager for open data in SKL" (SKL Swedish Association of Local Authorities and Regions, 2015).

The main ruling parties at least support the importance of open data and data transparency, as is visible at the forefront of the Swedish government information portal[5] (Sweden.se).

In terms of comparative politics, Swedish municipalities and regional authorities are known to own a lot of their creative processes. The topic of open data is more and more frequently politically backed, as by the Nya Moderaterna party last year:

"The County Council is sitting on a gold mine of information, the problem is that it often stays inside the doors,"

says Robin Karlsson (M).

"By making this information available, we can get more people to work together to make Sörmland Sweden's healthiest county".[6]

Sweden also has the Pirate Party, which although not growing in strength, has filled a niche of advocating intellectual property ultra-liberalism on a universal scale. In 2013, they took a hard line

[4] https://www.lantmateriet.se/sv/Nyheter-pa-Lantmateriet/lantmateriet-forenklar-villkoren-for-oppna-geodata/.

[5] https://sweden.se/society/openness-shapes-swedish-society/.

[6] http://www.moderat.se/pressmeddelande/det-oppna-landstinget-fran-vision-till-verklighet.

on an open data agreement by voting no to a watered down draft in favor of a superior one.

"Open data and free models are simply the best for both social interaction and the economy in general, according to many studies", says Amelia Andersdotter of the Pirate Party.

"None of the ongoing projects in the [European] Union will be helped by the agreement as it stands now. We might as well wait to solve problems with a second reading, rather to throw away a half-finished product into Union law".[7]

On the opposite spectrum, the ruling Social Democratic Party has supported thorough open data initiatives such as the ones undertaken by the City of Örebro.[8]

"Open data, among other things, created new opportunities in the e-business sector, while it has increased the accessibility and transparency of municipal citizens".[9]

The Folkpartiet has supported municipal efforts to open datasets, as for example in, Northern Stockholm Municipality, Sundbyberg.

3.1.6. *Non-profit*

The appearance of non-profits focused on open data is relatively new in Scandinavia and their true impact is yet to be gauged. A natural role for non-profits in the Swedish open data ecosystem might be to curate and market open data opportunities and support the innovation chains with effective matchmaking, and marketing of open data opportunities. The primary challenge for newcomers in an ecosystem is both finding a voice and finding a niche.

Another large non-profit is the Swedish Standards Institute. Its growth in both marketing and matchmaking has been slow, and it serves primarily as an index.

[7] http://www.mynewsdesk.com/se/pressreleases/baettre-vaenta-med-oeverenskommelse-om-oeppen-data-859105.

[8] www.orebro.se/psidata.

[9] http://www.socialdemokraterna.se/Webben-for-alla/Partidistrikt/orebrolan/Nyheter1/Pressmeddelande-Civilminister-Ardalan-Shekarabi-besoker-Orebro/.

On top of institutes and universities, small organizations are starting to appear which focus on finding niches inside this ecosystem.

Wikimedia and open knowledge Sweden are two examples of non-profits who engage in the ecosystem as innovation missionaries. Their mission is to market, promote, raise awareness, document, and foster interaction between parties in fields relating to open data and its role within open innovation and e-government. Both organizations function in the same way, the nuanced difference between their modus operandi being that Wikimedia also provides funding as an innovation investor and OKS is seeking to catalyze interactions within the diverse sections of the open knowledge ecosystem.

3.2. *Generating innovation*

Chesbrough suggests organizations generating innovation can be divided into four categories:

1. innovation explorers,
2. merchants,
3. architects, and
4. missionaries.

3.2.1. *Innovation explorers*

Innovation explorers focus on the discovery element, the element of research which was previously a part of the traditional corporate R&D center (Chesbrough, 2003). Chesbrough describes "interesting developments" occurring at public research institutions, including Sandia National Laboratories, Lawrence.

3.2.1.1. People as service users and government e-service administrators

The population using services is an innovation explorer in this case. The two way process of information portals, their usage and their

Table 4. Innovation explorers.

Organization name	URL	Open data exclusive	Years active	List category
Forum for Social Innovation in Sweden	http://www.socialinnovation.se/en/about/	No	Since 2008	Non-profit
SSC Space	http://www.sscspace.com/	No	Since 1972	Business
Meta Solutions	https://metasolutions.se/products/	No	Since 2011	Business

accessibility, especially in the realm of government e-services, provides a force of innovation explorers.

A good example is the very efficient combination of services from different agencies available at https://open.fk.se/ — a portal which was created by the Swedish Social Insurance Agency or *Försäkringskassan*, to render their collection of services more accessible to migrants and new arrivals. Despite not being complete, it is already attracting users at least as a data broker. The effort is born out of a one-stop idea, building a portal for migrants, a concept emerging out of the *Mötesplatsen för Nyanlända* project.

The parent agency, *Försäkringskassan*,[10] already has a comparatively large number of users for their e-services, and arguably is a pioneer for government e-services in Sweden.

There is a significant culture of e-government service usage in Sweden, but a lot less is done on citizen initiative.

The Swedish Migration Agency or *Migrationsverket*[11] provides extensive services, although many of them still require an office visit. Most other agencies in Sweden provide some form of services when transactions with citizens are at stake.

In this category, Sweden ranks high and it is in the citizen and public participation bracket where the attainment is lower.

[10] https://www.forsakringskassan.se/.

[11] http://www.migrationsverket.se/Privatpersoner.html.

Another very interesting example is the Open Stockholm Portal[12] which lists datasets. This is a service which actually has a growing number of participants, especially from explorers in the innovation chain.

3.2.1.2. People as citizens

Services and awareness campaigns: are they increasing citizen activeness?

In Sweden, there have been a moderate number of web-based awareness raising campaigns, but for the most part, the amount of participation is less than expected compared to the volume of information made available.

An interesting example is OpenAid, which seeks to render transparent all aid funding undertaken by the Swedish Government (such as directly from a Ministry) and Swedish Development Agency (SIDA). Although the site is extensively designed, with numerous data available which guides the public and public sectors, there have been indications that the citizen's involvement has been lower than initially expected, and instead the site is mainly used by parties which rely on funding, therefore falling short on an original intention to help shape development aid policy with the participation of the Swedish (and perhaps non-Swedish) public.

Open data impact conducted a case study on OpenAid.se, and it is visibly linked on the site's homepage. A very worthwhile read, a well-conducted study OD Impact, makes the following remark in their "key takeaways":

"A lack of clearly defined, high-level political commitments to publishing open data and enabling reuse can create major, but not insurmountable challenges. While Sweden now boasts such high-level commitments, throughout much of openaid.se's development, no such policy existed".

Furthermore, citizen input on development aid fell short as a priority in the construction of OpenAid, although it was officially

---

[12] http://dataportalen.stockholm.se/dataportalen/?loc=en.

a priority in the Swedish Government during the time of its construction, as OD Impact remarks:

"When Sida first launched OpenAid.se in 2011, user experience wasn't a key priority. As such, the site offers limited opportunities for citizen engagement or interaction. The only feedback channel is an opinion button which people use for reporting bugs. Usability is also a concern, and project titles often contain cryptic terms that are interpretable only to those who work internally on projects, making them difficult to understand for a wider audience. As such, there is a need to raise awareness among staff to ensure the accessibility of data with outside users".

It is our hypothesis that citizen participation in Sweden is caught in a closed circle of undernourished efforts. OpenAid illustrates how the lack of belief from policy makers that citizens would participate led to a bypassing of participatory mechanisms as an architectural priority. We would maintain that it is precisely a lack of resilient effort to instill a culture of participation and market such structured citizen-input opportunities alongside relevant information would prove effective.

3.2.2. *Innovation merchants*

Merchants also scout and function as explorers, but they reduce their viewpoint, aggressively patent and acquire intellectual property, and fervently sell and broker to third parties (Chesbrough, 2003).

Basically, this list encompasses companies which see market value in open data innovation and have bought and sold efforts within the Swedish ecosystem.

3.2.3. *Innovation architects*

Architects fill a sort of role of curator in a complex innovation ecosystem. They bring frameworks to the innovation value chain, and cut through the complex chaos, modularizing parts of the research so that different components can function together once divided

Table 5. Innovation merchants.

Organization name	URL	Open data exclusive	Years active	List category
STING	http://www.stockholminnovation.com/en/	No	2002	Business
Nemo-Q	http://www.nemoq.com/	No	Since 1978	Business
Sitevision	http://www.sitevision.se/	No		Business

within the innovation process. This allows for an industry to emerge around these frameworks of companies which can provide sections of the research along the lines of standards which allow for innovation within a greater innovation sphere (Chesbrough, 2003, p. 39).

Chesbrough gives the example of Boeing, which creates an overall design of an airplane, while other companies, such as GE for example, can use the standards set to complete the process with smaller components with a certain degree of innovative freedom.

Innovation architecture is complex, expansive and demands a rapid tempo. When collaborating on such complex endeavors, timing becomes even more essential. The very culture of innovation architects is contradictory to the "do-it-yourself" or other propriety models (Chesbrough, 2003). Architecture also has to function in a more complex profit environment, and when the creation process is divided, the profitability can be affected, rendering sustainability more challenging (Chesbrough, 2003).

KYAB can be described as an organization that fits this category. KYAB helps industries, businesses and property owners to reduce their energy consumption by combining measurement, visualization and counseling with people in mind. KYAB received funding of 786,580 SEK from Vinnova to develop APIs and tools for data sharing.

3.2.4. *Businesses as innovation architects*

Industry, especially the car industry providing a good example, is increasingly using open datasets both for open innovation and

Table 6. Innovation architects.

Organization name	URL	Open data exclusive	Years active	List category
KYAB	http://www.kyab.se/index.php/en/about	No	2006	Business
Bisnode	https://www.bisnode.se/	No	Since 1999	Business

marketing. In some cases, they are generating their own open datasets based on user experience inside, for example, a Volvo automobile, such as is possible via the "You Inside" app provided by the carmaker.

In this case,[13] it is perceived that "this information will help Volvo to design new and better vehicles according to customers' needs", which is both a potential innovation drive and marketing policy.

The biggest frequency of the intersection between open data and open innovation is naturally in the app market, which is trending in diverse industries ranging from airlines to pharmaceutical companies. An example here is Omega-Pharma.se collecting over-the-counter medicine usage via its apps[14] or SAS airlines trying to gather better data on customer preferences and needs while providing essential information via an app.[15]

STING, a Stockholm Business incubator/investor, has supported a number of open source and API projects to date since 2002.[16] It can be argued that this category functions as innovation architects, because of their ability to gather data and influence their innovation chain with external input.

3.2.5. *Innovation missionaries*

Actors who are driven by a cause or mission in their research and innovation function as missionaries. They are not after a lucrative goal, but instead seek to innovate for a cause greater than money or

[13] http://opensweden.net/examples-of-open-innovation-in-the-automotive-industry/.

[14] http://www.omega-pharma.se/.

[15] http://www.sas.se/en/travel-info/mobile-services/apps/.

[16] http://www.stockholminnovation.com/en/2014/10/02/open-call-e50000-develop-new-web-based-project-idea/.

Table 7. Organization's innovation category examples.

Organization name	URL	Open data exclusive	Years active	List category
Wikimedia Sweden	https://se.wikimedia.org/wiki/Huvudsida	Yes	~6 years	Non-profit
FORES	http://fores.se/about-fores/	No	~10 years	NGO
Open Knowledge Sweden	www.okfn.se	Yes	~3 years	NGO

direct material benefit (Chesbrough, 2003). Wikimedia Sverige and Open Knowledge Sweden are two non-profit organizations that promote and raise awareness, and work on open data-related projects. Nordic APIs (http://nordicapis.com) arrange several seminars per year in Stockholm to disseminate knowledge and information, which are designed to help organizations become more efficient via the use of APIs to come up to develop new services and solutions.

3.3. *Commercializing innovation*

The marketing of innovation is of importance, as metadata is to data. Chesbrough (2004) divides the subset into innovation marketers and one-stop centers.

3.3.1. *Small to medium enterprises*

There are conscious business ventures dependent on open datasets, and these extend to further than app manufacturers. Examples here are numerous — the *Stolen I Solen* app won plaudits in 2014 for using Stockholm open geographic data.[17] Another awarded app, LUP, used open data sources to identify optimal loading and unloading zones at different times of the day in Stockholm's dense traffic environment, providing address of zones near a delivery address.[18]

[17] http://www.liedman.net/2014/06/25/sunshine/.

[18] https://www.kth.se/innovation/nyheter/storslam-for-lup-i-open-stockholm-award-2014-1.483990.

Due to a healthy open data policy pursued regularly in recent years by the City of Stockholm, a vibrant number of apps is available based on city open data.[19]

Overall this sector is growing very rapidly in Sweden, and scores of small to medium enterprises (SMEs) start to use open data as part of their service and product development.

SMEs in Sweden can be categorized as innovation marketers.

3.3.2. *Innovation marketers*

Very often marketizing is not an activity exclusive of functions performed by other types of organizations involved in open innovation. The benchmark here is the drive to increase profit margins based on the promotion of both in house and external ideas. To be successful, a deep understanding of the subject matter and the market context as well as timing is necessary. Marketers for this reason are also good brokers for outside ideas (Chesbrough, 2003). Chesbrough gives an interesting example of Pfizer, a pharmaceutical company which currently has a catalogue with largely external origins (Chesbrough, 2003).

When a main activity involves a certain market knowledge, it positions an actor as an ideal discerner of ideas irrelevant of internal or external origin. Innovation itself is a beast in the ecosystem which is best nourished by a steady and continuous supply of fresh ideas, to reinforce or improve what is already there without a threshold (Chesbrough, 2003).

Open data innovation that does not automatically turn into a product needs marketing. These organizations give quite a bit of exposure to technologies and innovative initiatives through their astute marketing efforts.

3.3.3. *One-stop centers*

Innovation one-stop centers provide comprehensive products and services. They take the best ideas (from whatever source) and

[19] http://www.slowtravelstockholm.com/resources-practicalities/stockholms-best-apps/.

Table 8. Innovation marketer examples.

Organization name	URL	Open data exclusive	Years active	List category
World Favor	http://new.worldfavor.com/	No	Since 2010	Business
Dopter	http://dopter.se/	Yes	Since 2006	Business
Media Evolution	http://www.mediaevolution.se/en	No	Since 2008	Non-profit

deliver those offerings to their customers at a competitive price (Chesbrough, 2003).

A rare but growing breed of organizations or companies provides a truly integrated package, and these are called one-stop centers by Chesbrough (2003). They do not discriminate the origin of ideas but only their quality, while marketing and delivering these competitively to the appropriate markets Chesbrough (2003). According to Chesbrough, "Some companies focus on funding, generating or commercialisation of innovation some do all three" (Chesbrough, 2003).

In Sweden at present, we cannot identify any one-stop centers. There is only a multi-agency project to provide a "single" point of access for open data (oppnadata.se). This new portal already has 330 machine readable datasets available in its index and is built to grow exponentially but it does not fit one-stop centers.

3.4. *Non-recursive path*

As demonstrated in Fig. 1, West & Bogers describe the variable "which is the non-recursive path which includes feedback loops, reciprocal interactions with co-creation partners, and integration with external innovation networks and communities" (West & Bogers, 2011).

The very nature of the open innovation process with open data makes it a very collaborative field. In most stages of innovation, feedback and interactive cocreation are standards. If you take three main open data portals described earlier, oppnadata.se, kolada.se or openaid.se, they were created in partnership between organizations and involved a phase of broader consultation.

e-Government Project Manager Pernilla Näsfors said, “More help using open data can be available to more actor to build climate smart services or examine how municipalities can fit in this service” (Näsfors, 2016).

The active citizen like the supportive politician both influence the process and help enable it as active users and promoters.

3.4.1. *Suppliers and partners*

Brokerage and generic joint investment in open data are still two grey areas. Although they materialize in certain instances, the biggest supplier is by and large the Swedish state.

Sveriges Komuner och Landsting (SKL) has taken the mantle of marketing open data and of promoting it through several partnerships.[20] Most notably, SKL partnered with the Council on Local Government Analysis (RKA) and the Swedish Government to create and administer Kolada.se, a linked open data portal which collates and makes accessible over 3000 key data figures from Swedish municipalities via API.[21]

One of the biggest efforts made by a cluster of organizations including Vinnova is open data dot SE, or http://oppnadata.se/about. While still under collaborative construction, this effort is a milestone for standardized data dissemination in Sweden, and its clean, sleek build could serve as a model for similar ventures internationally:

“It will serve as a catalogue, in the sense that it will identify the data sources of the data owners (the portal is not a custodian of data). The goal is that the portal should be simple to use, with clear licenses and conditions, provide support for that data can be re-used to

---

[20] http://skl.se/naringslivarbetedigitalisering/digitalisering/digitaldelaktighetoppenhet/oppnadata/arkivoppnadata/kommuneroppnaruppsinadatabaser.4532.html.

[21] https://www.kolada.se/?_p=index/about.

innovate and enable the sharing of resources and solutions. This platform is a first version, which will primarily be to evaluate the technical solution. Therefore, the number of data sources is limited for the moment. The platform is clear and very development remains — we wish consultation and proposals for the next step. In late January we will organize a discussion on the evaluation and further development".[22]

The constituent document, which preceded this build of the oppnadata.se portal, reveals a lack of identifiable suppliers on a large enough scale to support a true open data initiative. It also seeks to address the issue of Sweden falling behind in participation despite having an exemplary set of legal preconditions and basic infrastructure.

Statistics Sweden (SCB) makes all their statistics available via an API. This early step means there is a lot of potential for innovation around the supply of their key statistic product.

The website suggests "The ambition of the Swedish Government's digital agenda is to improve conditions for the reuse of public information from government agencies for both commercial and non-profit purposes. This is also regulated in the PSI Directive and the INSPIRE directive from the EU".[23] The limitation is that the use API is also limited based on the caller's IP address. The caller can make a maximum of 10 calls within 10 seconds.

3.4.2. *Meetups, hackathons and Hack for Sweden*

Hack for Sweden[24] is a unique partnership of 17 Swedish agencies and organizations, and is growing every year. The partner organizations contribute data, finances and other resources during the actual hack weekend. Since 2014, an annual competition is held, where students, developers, data journalists, innovators and government agencies get together to see who can create the best digital services

[22] http://oppnadata.se/about.
[23] http://www.scb.se/en_/About-us/Open-data-API/.
[24] http://hackforsweden.se/.

using open data. Entries are evaluated based on four categories: best visualization, best business value, the best public benefit and cost efficiency.

It is interesting to look at Hack for Sweden partners as a mini-ecosystem because all 23 are different public bodies with different datasets, which together give access to quite a comprehensive opened and unopened pool of data. Yet, it is beneficial for these organization to join forces and attempt to innovate within a common realm together with external catalysts in order to apply economies of scale, and in order to get a more vibrant innovation sphere. They attain cost cutting, quality of output, quality of outreach and marketing, access to more resources and good practices, through collaborative work.

Nordic APIs[25] is an important organization to promote the use of APIs to help average API practitioners to enhance their activity and to help understand the very role of APIs in open data innovation.[26] The collaborative business model of this organization, which relies on a regular engagement in partnerships, makes it a prime example of a capable enabler within the innovation chain in Sweden.

Open knowledge Sweden and Wikimedia organize unofficial meetups which bring actors from a wide array of sectors together to discuss and develop collaborative approaches to open data and e-government projects. They are enables as well as innovation missionaries within the Swedish ecosystem.

4. Discussion and Conclusion

The world's first Freedom of Information Act was proposed by thinker and politician Anders Chydenius and adopted by the Swedish parliament in 1766 (Björkstrand & Mustonen, 2006). The key successes of the 1766 Act were the abolishment of political censorship and the gaining of public access to government documents (Manninen, 2006).

---

[25] http://nordicapis.com/create-with-us/.

[26] http://nordicapis.com/api-insights/platforms/.

These early milestones, albeit impressive, have fostered a certain degree of excess comfort and arguably, have contributed to a lesser degree of civic engagement. Sweden can still attain a greater degree of maturity, and a good amount of market actors engage in opening datasets through old-school legal obligations rather than an actual drive for innovation.

For this reason also, there is an undeveloped segment of actors and a large number of passive actors who are not fulfilling their mandates in Sweden. Furthermore, a critical mass of actors are private individuals and consultants, responding to the funding environment, rather than established organizations and companies operating on more sophisticated business models.

As a topic, open data innovation does not get significant prioritization outside of big data or even private proprietary data endeavors. In a country where databases are still frequently sold and where transparent salary and address information are sources for closed-source profit, there is a resistance to "giving away" traditional profit chains.

Also, Sweden is a heavily government-financed country, which translated to a (budgetary and working hour) fear from public officials that they must drag the entire movement of their own accord. With the highly proprietary approach to data within the business and non-governmental sectors, this combination creates an occasional apathy toward "open first, profit later" approaches.

Pricewaterhouse Coopers evaluated open data usage in Sweden in 2014 and stated the following:

"With the Swedish tradition of transparency in public administration and high level of technological maturity, Sweden has achieved progress working with open data" (Pricewaterhouse Coopers, 2014).

However, their study went on to conclude

"According to E-delegation recent survey has shown more than two-third of the authorities has not delegated the responsibility to anybody to work with open data and about half of them have no planed budget for open data activities, said Magnus Kolsjö, senior manager and director of open data at PwC" (Pricewaterhouse Coopers, 2014).

This context has slowed and fragments the development of modern/vibrant e-government initiatives in Sweden, despite the fertile legal, social and democratic context. Municipalities and regional governments (a key echelon in Swedish governance) engage very selectively in eGovernment projects, with instances of public question or public participation being as rare as instances of web-based public opinion gauging. Direct query is rare, with a culture of hiring consultancies and conducting assessments and garnering input tending toward not being truly accessible.

While the individual initiatives and even multiple instances of stalwart government support are to be factored in, there remains a lack of mainstreaming of the subject by political parties and the press. While individual politicians act as enablers and important instances of open data promotion are visible in government strategies and documentation, the link between innovation and data is not made in policy agendas, and thus very rarely makes it in the mainstream rhetoric.

While there are clear examples of individual political support, and lacing of important strategy and policy documents with open data and open innovation, in general both do not take a rightfully strong place in government rhetoric in general, nor do they capture the attention of the mass media, or broader-public marketers.

Sweden seems only on the verge of including and therefore empowering the citizen as an active open data user. At present APIs and open data are still very much dismissed as a geek project which might be important, but is not key. Knowledge of visualization, points of access, data depositories and indexes, and even basic knowledge of what open data is are still really low in Sweden. The paradox remains that even though Sweden has an exceptional funding and professional environment around open data, a very old and established legal precedent creating an excellent precondition for e-government, there is a lack of prioritization around supporting it as a platform.

Lack of one-stop centers and proper open data business models can also be two of the explanations of not having an innovative open data ecosystem in Sweden. New business models should be explored and tested with different stakeholders in order to create innovative open data applications.

We also noticed that Chesbrough's (2003) description of innovation modes is limited to address the Swedish open data ecosystem. Combined with West & Bogers (2014), both frameworks should be further analyzed and tested for other open data ecosystems.

References

Attard, J., Orlandi, F., Scerri, S. & Auer, S. (2015). A systematic review of open government data initiatives. *Government Information Quarterly*, *32*(4), 399–418. Retrieved from http://doi.org/10.5281/zenodo.18592.

Bonina, C. M. (2013). New business models and the value of open data: Definitions, challenges and opportunities. *RCUK Digital Economy Theme*. Retrieved from http://www.nemode.ac.uk/wp-content/uploads/2013/11/Bonina-Opendata-Report-FINAL.pdf.

Broster, D., Misuraca, G. & Bacigalupo, M. (2011). Lifting off towards open government: A report from the EU Belgian presidency conference. *European Journal of ePractise*, *12*, 53–65.

Chesbrough, H. W. (2003). The Era of Open Innovation. *MIT Sloan Management Review*, *44*(4), 26–32. Retrieved from http://dialnet.unirioja.es/servlet/articulo?codigo=2316408.

Freeman, R. E. (2004). The stakeholder approach revisited. *Zeitschrift Für Wirtschafts-Und Unternehmensethik*, *5*(3), 228–241. Retrieved from http://doi.org/10.3763/jsfi.2010.0008.

Gauld, R., Goldfinch, S. & Horsburgh, S. (2010). Do they want it? Do they use it? The "Demand-Side" of e-Government in Australia and New Zealand. *Government Information Quarterly*, *27*(2), 177–186. Retrieved from http://doi.org/10.1016/j.giq.2009.12.002.

Gurin, J. (2014). An Infographic: Big Data is BIG. Open Data is REVOLUTIONARY. McGraw-Hill BusinessBlog. Retrieved from http://www.mcgrawhillprofessionalbusinessblog.com/2014/02/18/an-infographic-big-data-is-big-open-data-is-revolutionary/. Accessed 14 April 2016.

Immonen, A., Palviainen, M. & Ovaska, E. (2014). Requirements of an Open Data Based Business Ecosystem. *IEEE Access*, *2*, 88–103. Retrieved from http://doi.org/10.1109/ACCESS.2014.2302872.

Nam, T. (2012). Citizens' attitudes toward Open Government and Government 2.0. *International Review of Administrative Sciences*, *78*(2), 346–368. Retrieved from http://doi.org/10.1177/0020852312438783.

Näsfors, P. (2016). Retrieved from http://www.y2s.se/2016/04/04/1000-km-hallbar-vandring-fran-ystad-till-stockholm-om-oppna-data-och-fns-globala.

Ohemeng, F. L. K. & Ofosu-Adarkwa, K. (2015). One way traffic: The open data initiative project and the need for an effective demand side initiative in Ghana. *Government Information Quarterly*, *32*(4), 419–428. Retrieved from http://doi.org/10.1016/j.giq.2015.07.005.

Poikola, A., Kola, P. & Hintikka, K. A. (2010). *Public Data — An Introduction to Opening Information Resources*. Helsinki, Finland: Edita Prima Oy. Retrived from: https://julkaisut.valtioneuvosto.fi/bitstream/handle/10024/78201/Public_data_-_an_introduction_to_opening_information_resources.pdf?sequence=1.

Pricewaterhouse Coopers (2014). *Öppna data i Sverige*. Retrieved from www.pwc.se.

Robinson, D. G., Yu, H., Zeller, W. P. & Felten, E. W. (2009). Government data and the invisible hand. *Yale Journal of Law & Technology, 11*, 160–175. Retrieved from http://doi.org/10.1177/0020852312438783.

Sandoval-Almazan, R. & Gil-Garcia, R. J. (2015). Towards an integrative assessment of open government: Proposing conceptual lenses and practical components. *Journal of Organizational Computing and Electronic Commerce*, *9392*. Retrieved from http://doi.org/10.1080/10919392.2015.1125190.

SKL Swedish Association of Local Authorities and Regions (2015). *Nytt verktyg for publicering av lankadoppnadata*. Retrived from https://skl.se/naringslivarbete digitalisering/digitalisering/informationsforsorjningochdigitalinfrastruktur/oppnadata/sklsverktygforattpubliceraoppnadata.7794.html.

West, J. & Bogers, M. (2011). Profiting from external innovation: A review of research on open innovation. In *9th International Open and User Innovation Workshop*, 1–47. Retrieved from http://doi.org/10.2139/ssrn.1949520.

West, J. & Bogers, M. (2014). Leveraging external sources of innovation: A review of research on open innovation. *Journal of Product Innovation Management*, *31*(4), 814–831. Retrieved from http://doi.org/10.1111/jpim.12125.

Yu, H. & Robinson, D. G. (2012). The new ambiguity of “open government”. *59 UCLA Law Review Discourse*, *178*, 178–208. Retrieved from http://papers.ssrn.com/sol3/papers.cfm?abstract_id=2012489.

Chapter 4

Digital Transformation via Open Data in Insurance

Bernardo Nicoletti

Universita' di TorVergata, Rome, Italy
Temple University, Rome Campus, Italy

"Chiunque desideri costante successo deve cambiare la sua condotta con i tempi".

Niccolo' Machiavelli

"Whosoever *desires constant success must change his conduct with the times*".

This chapter aims to present the usefulness of using open data in insurance services in support of their digital transformation. It underlines the ways in which public administration and private companies could use them. This chapter underlines the limitations but also the benefits of such an approach.

The chapter defines a digital transformation strategy for insurance companies. It considers the definition of a business model. On this aspect, this chapter uses a modified business model canvas and details all its components. This chapter also defines the need of governing the digital transformation and some of the key organizational profiles in connection with the use of open data.

The second part of this chapter analyzes the potential development of open data in connection with insurance companies in the future. This chapter underlines the importance that the blockchain technology might have in the future for moving toward digital insurances. At the same time, this development might underline even more the relevance of open data in insurance.

The approach is innovative especially in the linkage of insurance with open data. This innovation stresses the importance of open data and blockchain for its future uses in insurance activities.

Academicians and practitioners will find this chapter relevant to their studies and to their practice.

1. Introduction

Over the last few decades, the financial services sector has faced many challenges, resulting from multiple sources: financial and economic crisis, (de-, re-) regulation; dominant role of information and communication technologies, service bundling; and changes in customer preferences. This has hit financial institutions and induced stronger needs for innovation. This chapter addresses the possibility of using new solutions to support innovation in insurance companies as opportunities to overcome some of the challenges.

The last few years have seen a drastic move to transform digitally every activity. The wave has affected also traditional businesses like insurance companies. Insurance is typically considered one of the functions within financial services where the adoption of innovation has been the slowest.[1] Over the past decades, many insurance companies have gradually adopted many innovative practices such as digital channels and process automation. This has been especially true in personal lines of business while large commercial lines have continued to focus on establishing a personal touch across the value chain. The objective of this digital transformation in all types of

[1] http://www3.weforum.org/docs/WEF_The_future__of_financial_services.pdf. Accessed 16 April 2016.

insurances is to improve the customer experience and at the same time improve the internal value stream.

One of the pillars of digital transformation in general and in insurance, in particular, has been an extensive use of the five main technological advances: Internet of Things, Cloud Computing, Artificial Intelligence and Robotics, Mobility, and data analytics (Nicoletti, 2016). Particularly interesting has been the latter aimed at optimizing an increasing volume of data, exploiting them at a higher velocity, using a variety of data to add value to the customers and reduce vulnerability (the five *V*s).

The data used until now in insurance companies have been mostly internal data. Only in a few cases, some insurance companies have started to use data external to the company mainly from social networks. The extensive use of open data is a worthwhile investment. It will pay off more and more, faster and faster as new data is made available in processable formats. Open data is a very interesting source of information, not only for retail insurance but also for commercial or wholesale insurance. Some initial uses have started. Until now, there is not a comprehensive analysis of how open data could be used by insurance companies. The definition of such a model is the objective of this chapter. To analyze this opportunity, the chapter starts defining what a digital transformation is and later moves to the business model of open data in insurance. This chapter closes with a glimpse of the future.

This chapter would be of interest to academicians but also to practitioners in the business world. The reason is that if incumbent insurance companies do not exploit this data treasure and do what could be called an open data monetization, somebody else will do. More and more insurtech companies (mainly start-ups combining insurance with advanced financial services) are launched and some of them exploit and will more and more exploit the open data treasury.

2. The Digital Transformation

Some believe that digital transformation is simply a matter of using digital technologies to sell and service clients more effectively, more efficiently, and in a more customized way. There are other interpretations of what digital transformation is:

- A new application of digital marketing initiatives;
- A matter of using technology to drive business process innovation;
- Nothing less than to be the Uber of taxi or the Airbnb of hoteling;
- and more.

In order to analyze digital transformation, it might be interesting to refer to a sentence of Rudyard Kipling in his book, *The Little Elephant* (Kipling, 2013). This author presented the so-called five *Ws* and one *H*. According to Kipling, for a problem to be considered complete, it must answer five questions starting with an interrogative word. In the case of digital transformation, this would mean to answer the following questions:

- *Why* digitally transform the organization?
- For *Whom* to do it?
- *What* is the product it should aim to provide?
- *Where* can it take place?
- *When* can it take place?

Some authors have added a sixth question to the list of Kipling: "How can it be done?"

Using this framework, the broad questions executives should be asking in implementing a digital transformation in an insurance company are then as follows:

- Why: The reason to do a digital transformation is to improve the business from an effectiveness, efficiency, ethical and economic point of view. The real nature and impact of the digital transformation of an established industry are not always obvious. For example, Uber may have a dramatic impact beyond the taxi industry in the years to come. By making personal transport an affordable service commodity, it could eat away at the edges of the car and auto insurance industries. In other words, the shared economy could have some drastic impacts also on insurance.

- Which are the best insurance companies across the spectrum of digital enablement? What can the executives learn from them about the future of the industry and the business? Organizations must understand how customers behave rather than simply looking at direct competitors. Remaining relevant is simply not a matter of creating an app or smartening up a website. It is essential to find ways to use customer data to create more meaningful and relevant customer experiences at every contact, be it physical or virtual.
- Who? Digital transformation requires a change on how institutions understand and engage with customers using digital tools and channels. This is an imperative and no longer up for debate. Unless this is done, nothing else is possible. This approach has the advantage of being realistic and manageable to implement.
- Where should the organization change defending and extending market share, grow profits and ensure relevance as digital technology evolves in the years to come? Which channels should be considered and integrated?
- When should an insurance company invest in innovative financial technologies? The simple answer "always" is in contrast with the realities of the possibilities of any single insurance company. By looking closely at competitors and the technology landscape, executives need to intercept low signals on how emerging technologies and disruptive rivals could attack their market shares. They need to create, deploy and manage the strategies necessary to protect their market share and identify ways to expand into new service/market domains using digital transformation.
- How exactly could digital technology change the experience the company's customers have and the way that existing, emerging and potential competitors use and innovate ways to do business?

The real problem is not so much a definition of what digital transformation is, but what should be the strategy in front of these challenges/opportunities and how should an organization be aligned within a digital transformation vision? What often executives do not know is how to bring about the changes that will help their organizations to be profitable, sustainable and competitive in an era of disruptive change.

The next step is how to implement the new strategy. The answer is not the same for every business. Some businesses will have visionary leadership, agile processes, innovative cultures, open workforces with digital skills and modern technology platforms, so they are able to embrace digital transformation more wholeheartedly.

In dealing with the implementation of a digital transformation, it is important to refer to a model of innovation. The approach to digital transformation should be holistic. To approach this challenge, it is possible to refer to a combination of the Chandler model of connecting Strategy and Structure (Chandler, 1990) and the Leavitt Diamond Model (Leavitt, 1965) by considering the four connected variables:

- Structure (organization);
- Processes;
- Technology; and
- Persons.

An example of this approach applied to a Digital Transformation Strategy is in Fig. 1 (Nicoletti, 2016).

Insurance companies might be encumbered by conservative leadership, legacy technology, regulation, siloed processes and non-receptive workforces. They will need to look at the available assets — data, customers, resources and channels — (internal and external to the company) and find ways to put them to work in a digital world. In some cases, they might need to launch new products, accept new business models or innovation groups to fast track their digital programs.

For the digital transformation to be sustainable, it is important to look at the three *P*s (Nicoletti, 2014):

- Products: The definition of services to be offered to the customers of the organization is really essential;

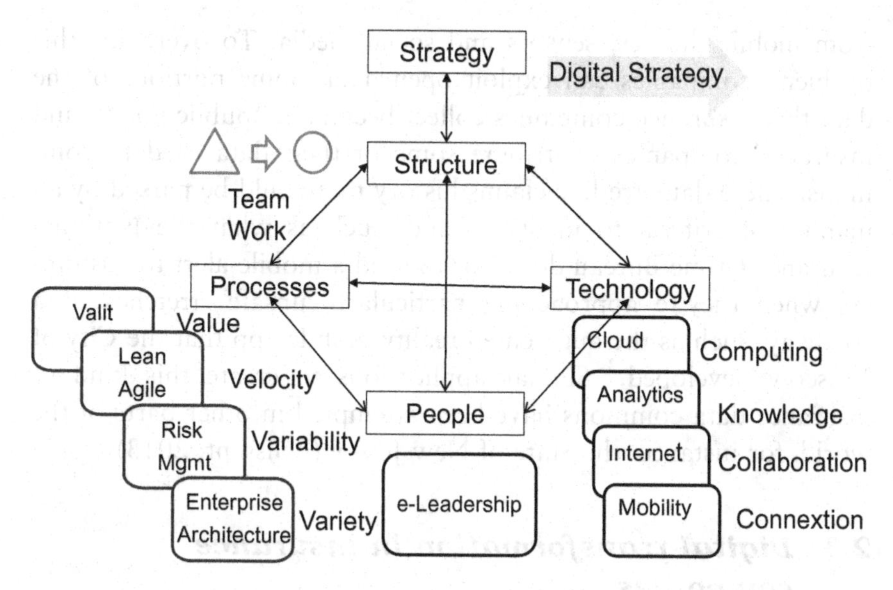

Fig. 1. A model for a Digital Transformation Strategy (Nicoletti, 2016)

- Processes: Processes are closely connected with the introduction of new products. The two goes together in an innovation;
- People: Finally yet importantly, the innovation in insurance must be deployed and delivered by people to people. To them, the executives should devote a very strong consideration and help individuate and foster the talents.

So ultimately, underpinning the organization's ability to perform a digital transformation lies in its ability to create a vision, define a plan, organize and make it real.

Insurance companies should also contemplate innovations through the power of data. For the last few years, Forrester has reported that data analytics has been the biggest investment boosts in insurance companies, year over year (Belissent, 2013). This is because companies recognize that data is a strategic asset for the business than the premium revenue streams.

On the other side, customer privacy is a huge issue in terms of how insurance companies are using all that behavioral data collected from mobile devices, sensors and social media. To overcome this problem, companies can exploit open data. Some portions of the data that insurance companies collect become a "public good" and insurance companies contribute some of their data to data commons. These data are like claims history that could be parsed by all manner of criteria to identify fraud, such as what the National Insurance Crime Bureau does,[2] or to send a mobile alert to customers when they're approaching particularly unsafe stretches of a roadway, such as the augmented reality mobile app that the City of Moscow developed.[3] Similar applications to create this kind of insurance data commons have been attempted in other parts of the world, for instance, the State of New Jersey (Belissent, 2013).

2.1. *Digital transformation in insurance companies*

Most insurance companies are reluctant to disrupt their own industries. Their concern is often because of the fear of cannibalizing their customer base or eroding their own margins. Many executives prefer to make minor changes to their business with digital technologies rather than to innovate their business models in a fundamental way. Clayton Christensen (2013) suggests that a new business, outside of the current business, is often the best way to have the better of two worlds. It means an organization can become more customer centric by using data and technology well in its current business, while at the same time experimenting with more disruptive solutions enabled by technology in separate offspring.

The recommendation is that executives should start by discussing their business's four Cs — context, customers, challenges and costs, and competitors — so that they can have a clear view of how digital solution, technologies and customer experiences can affect their organizations in the years to come.

[2] https://www.nicb.org/. Accessed 16 April 2016.

[3] http://designtaxi.com/news/350327/To-Curb-Traffic-Accidents-a-Macabre-Augmented-Reality-App/. Accessed 16 April 2016.

In the past, it has not been possible to design, sell, service products and services customized to individual customers (market segment of one). As the wealth of next generation of technologies makes this possible, the information architecture of life, health, property and casualty insurance companies will change drastically. Digital transformation can make all this possible and arrive at personal mass insurance policies and products. One interesting solution in this direction is based on open data.

3. The Open Data

Open data is data that is held openly and is free to access, use, reuse, redistribute and so on.[4] There are many sources of open data; for example, public administration data, weather reports, social networks and sporting outcome reports on news services. The open data movement to make available the access to public (and other) information is relatively new. It is acting as a powerful, emerging force. The overall intention is to make personal, local, regional and national data (and particularly publicly acquired data) available in a form that allows for direct processing using software tools, for example, for the purposes of cross tabulation, visualization, mapping and so on (Gurstein, 2011).

The idea is that public (and other) data, whether collected directly as part of a census collection or indirectly as a secondary output of other activities (crime or accident statistics, for example, but also social networks or similar) should be available in electronic form and accessible via the Web. There are significant initiatives in this area underway all over the world and as part of a wide variety of not-for-profit initiatives.

A certain number of governments have agreed to commit to moving toward open data as a matter of policy. For example, all the data that the British Government collects, which is not considered essential to national or public security, will be made open too.[5]

---

[4] Wikipedia, at http://en.wikipedia.org/wiki/Open_Data. Accessed 12 April 2016.

[5] http://www.insurancesoftware.co.uk/open-data-insurance-brokerage/. Accessed 13 April 2016.

In order to commit to open data, most data sources must be anonymized. This means any personally identifiable information should be masked or scrambled. This may become problematic in the long term. As more and more people put their personal data in the web or mobile sites, there might be a possibility to rebuild the anonymized parts of the data by comparing the datasets to other sources. With this precaution in mind, there are many data available to insurance companies that are freely available. Examples are statistics relating to traffic, health, crime, natural disasters and so on.

Freely available does not mean that they are ready to be used. It is necessary to select them, extract, transform and load into a private repository. This is not an easy task. It requires resources and special skills. A part of this chapter is devoted to deal in detail with this aspect essential to get the maximum benefits from open data.

4. The Exploitation of Open Data in Insurance

The insurance organizations are on the verge of some dramatic changes in their characteristics. In a more competitive market, insurance companies are relying more and more on innovation as a competitive advantage. According to a Deloitte study, financial services and insurance are the third most important sector that can use most open government data, after the obvious public administration and defense themselves and the information and communication data sectors (Deloitte, 2012).

Worldwide, there are millions of public available databases and sites. They are open for use. Other insurance companies can tap into the same data. It is essential to have the capability to select the appropriate data and use them in a timely, efficient, effective, economic and ethical way.

Some examples of open data available in a certain number of countries are as follows:

- Public Health: Some infringements to public health by businesses, like restaurants, are published and they can provide valuable information for fraud detection teams.

- Property: The land registry releases all the data on the price paid for a property, allowing you to track trends and identify new opportunities. If you add flood data to this, then you can start to create some interesting insights. This type of data is updated regularly so it is possible to stay ahead of the market by keeping them fresh.
- Roads: Road accidents, which involve the police, are captured and a new dataset is released periodically in many countries. This includes data on the weather, causalities and vehicles involved. Merge this in with the average number of penalty points issued in a postcode sector as released by the driver and vehicle licensing agencies and these data can give extremely useful insights on the risks of a specific prospect or of a territory.

The main difficulty is the fact that there is no central repository from which to select open data useful to the insurance companies. Each party, be it a central or local government entity or a private association or a social network can decide its own format and its own way to organize data. It is also necessary to govern the process and evaluate the benefits of investing resources and time to use open data. The business case is not easy to prepare especially on the side of the benefits. This will be another aspect discussed later in this chapter.

To defend their markets, insurance companies must build new business models that focus on adding value to the customers also through digital transformations. They should apply the capabilities of new solutions to improve the ways they assess risk and operate their businesses. The biggest winners will be insurance companies with the foresight to identify new game-changing innovation that may not be available for immediate utilization (for instance, in sales) but could have a significant long-term impact on the industry.

Insurance companies and brokerages can exploit open data in order to provide more tailored services and to estimate risk with greater accuracy. The innovation can be of many types. In order to analyze the possible innovation, this chapter suggests examining the Business Model Canvas (BMC) (Osterwalder & Pigneur, 2010) and see how each of its components could be innovated using open data. This analysis is similar to the one done on Big Data by Morabito (2015).

4.1. *The business model canvas*

A business model is a description of "how an organization creates, delivers, and captures value" (Osterwalder & Pigneur, 2010). An organization can be a single entity or a collection of entities working together to deliver a product or service that creates value for a customer.

The BMC is a convenient visual framework for analyzing a business model. The BMC is a poster format chart that enables capturing and discussing nine elements of a business model. The nine elements can be visualized in a canvas or a cartoon (see Fig. 2). This chapter uses this model for analyzing the use of open data in insurance. The components of the BMC, revised to adapt to the digital transformation, are:

- Market: Three aspects are important. They can be labeled the three Cs — the target customers, the competitors and the compliance necessary to respect. Who are the most important customers?
- Value Propositions: Which are the products and/or services? What value the company delivers to a customer in a given segment? What needs does it satisfy?
- Channels: How to reach each customer segment? What is easiest for the customer? How to integrate channels in an omnichannel approach (Cummins, Peltier & Dixon, 2016)?
- Customer Relationships: Which will be the customer experience? How to build and maintain this experience? How do they fit effectively in both the customer's world and our own?
- Revenue Streams: For what will the customers be willing to pay? How would they prefer to pay?
- Key Resources: Which resources are essential to deliver the company Value Propositions through the Channels and improve the Customer Experience? Which organization should be set up?

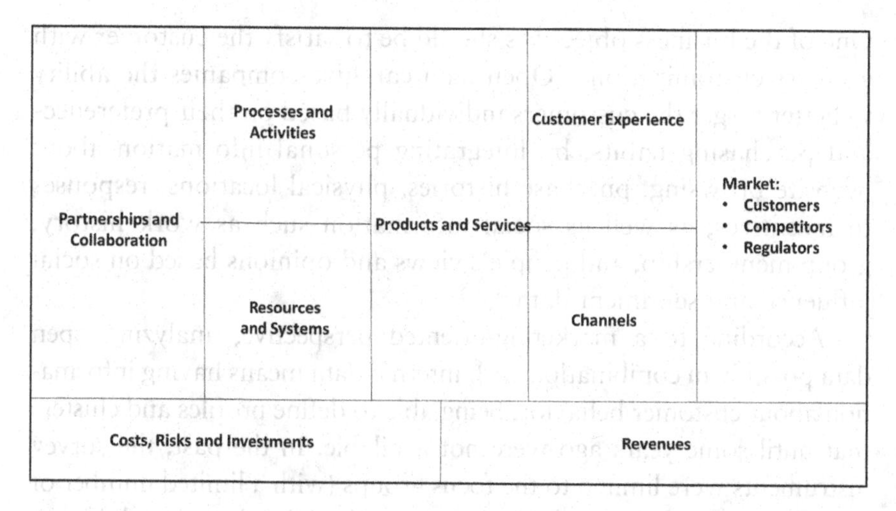

Fig. 2. The BMC (Adapted by the author from Osterwalder & Pigneur, 2010)

- Key Processes and Systems: Which are the most important activities to make the business work?
- Key Partnerships and Collaborations: Who are the Key Partners and Vendors? Which Key Resources do they provide and which Key Processes and Systems can they support? What is in it for them? What relationship should be with them?
- Cost and Risks: Which are the finance (costs and investments) implied by the Business Model? Which is largest? Which costs are fixed and which are variable? Which are the main cost drivers?

It is important to consider how the innovation could be implemented in these different components. All these types of innovation are important. Actually, there are innovations that can cover the full or part of the list. Sections 4.1.1–4.1.9 analyze the case of the insurance companies.

4.1.1. *Market (3 Cs: the target customers, the competitors and compliance)*

In the business, the primary objective is to identify the target customers in order to find the needs to satisfy with the product or service.

One of the business objectives should be to satisfy the customer with a "mass customization". Open data can give companies the ability to better target the customers individually based on their preferences and purchasing habits, by integrating personal information about website browsing, purchase histories, physical locations, responses to incentives, as well as social information such as work history, group membership, and people's views and opinions based on social influence and sentiment data.

According to a marketing-oriented perspective, analyzing open data possibly in combination with internal data means having information about customer behavior, being able to define profiles and clusters that until some years ago were not available. In the past, the survey instruments were limited to the focus groups (with a limited number of participants) and surveys. Market research today has the ability to query thousands of people using social networks, with the possibility to do web monitoring of communities, blogs and site searching.

From the point of view of the insurance companies, it is essential to have a way to "Know your Customer" (KyC). KyC is important from several points of view: risk management, marketing and finance. It is possible to use data analytics to support KyC. This is the process of examining large datasets containing a variety of data types — for instance, open data external to the companies — to uncover hidden patterns, unknown correlations, market trends, customer preferences and other information. In this way, it would be possible to provide very personalized insurance services.

In open data, a multitude of new types of data is readily available. These include:

- Activity-based Data: for instance, web sites tracking information, query and purchase histories, mobile data, responses to incentives or requests of information;
- Social Network Profiles: for instance, work history, group membership, hobbies, travel;
- Social Influence and Sentiment Data: for instance, product and company associations (including, likes or follows), online comments and reviews, customer service records.

This data explosion enables the definition of increasingly finer segments. These microsegments enable ever finer targeting of content, offers, products and services, which can deliver real and substantial returns.[6]

While open data is discussed primarily as a business tool, the customers themselves can use data to search and optimally procure products and services.

4.1.2. *Value propositions* (*Products and/or services*)

A product or a service is essentially a value proposition to the customer. There are many opportunities for insurance companies in exploiting open data to design their offers. Customers have a strong interest in insurance companies offering innovative products that apply new capabilities and are conveniently deployed. Insurance companies need to add value to their customers by addressing the customer experiences and operate much more effectively, efficiently and especially economical. This standing is not easy for incumbent insurance companies. They will need to pick up challenges such as

- Define and create innovative products and services that closely match their customers' lifestyle and preferences individually. In other words, insurance companies should aim to offer "mass private insurance";
- Price products more effectively and competitively;
- Gain better insight, process claims efficiently; and
- Minimize risk for underwriting and fraud detection.

Thanks to open data, insurance companies are able to collect information on these aspects and assess risks in completely improved ways with respect to what has been possible traditionally. This has the potential to radically reshape product propositions and reduce the size of risks. Property and casualty insurance is likely to see the

[6]Offsey, S. (2014). Retrieved from http://marketbuildr.com/blog/segmentation-in-the-age-of-big-data/. Accessed 12 April 2016.

biggest long-run impact from the technology transformation as it moves from actuarial risk assessment using statistical techniques to structural risk modeling based also on real-time observations of phenomena such as statistics on car accidents. Insurance companies that take hold of these opportunities can become the leaders in the markets.

Data made available free of charge from public bodies can be used in support of insurance companies in designing their products and services. This is particularly true for data that have broader potential value including mapping, meteorological, legal, traffic, financial and economic data. Much of this raw data can be used for or integrated into new data-enabled products, apps and services that individuals use on a daily basis, such as car navigation systems, weather forecasts or financial and insurance services.[7] According to a Finnish study, businesses that reuse geographical data grew 15% more per annum in countries where governments released such information freely, compared with countries that price such information in order to recover production costs (Ubaldi, 2013).

However, raw data is not of much use unless it is analyzed for a purpose. Insurance companies can develop higher value open data analytics services by recruiting data scientists to mine, analyze and synthesize data into consultancy services.

4.1.3. *Channels*

Expansion of the telematics insurance models through platforms (such as the social networks) will create channels for insurance companies to better understand their customers and engage more closely with them. Social networks are platforms that use the website or other technologies to allow different parties to communicate with each other and share information, resources and so on. By using open data in social networks, insurance companies can detect signals such as somebody is either buying a new item, be it a car or an apartment or similar, or is in search of a convenient insurance

[7] See examples in http://cloudmade.com/solutions/. Accessed 13 April 2016.

company. There are insurance companies that use Facebook as the primary acquisition channel.[8]

It is important to consider in the open data analytics results as well as the structured and unstructured content coming from social networks. The social network can also be an extremely important channel for promotion and advertising of the insurance products.

4.1.4. *Customer experience*

An open data innovation requires some drastic changes in the organization of the insurance companies. Insurance companies need to rethink completely their customer engagement processes. Customers' overall digital experience with insurance companies lags that of other industries. This is true particularly when it comes to "moments of truth" such as paying claims. As customers continue to integrate digital experiences into their lives, they expect these experiences, as well as their relationship with insurance companies, to become more direct, simple, seamless and effective. This is especially true in unpleasant moments of their life, such as when they need to file a claim.

The ways that insurance companies need to pursue these challenges is through innovative projects. This is not easy since traditionally insurance companies have relied on conservation rather than on innovation.

Profiling customers utilizing open data capabilities can increase sales with much information about customers obtained through open data, companies can improve customer service to increase customer intimacy and loyalty.

4.1.5. *Revenue*

Pricing is an important aspect that can benefit from the use of open data. An example is open data use in underwriting to set pricing on policies. Insurers can use public shared data to inform pricing decisions

---

[8] http://www.celent.com/system/files/the_hundred_percent_digital_insurer.pdf. Accessed 16 April 2016.

and aid in fraud detection, as researchers from Celent predict it.[9] Insurers could use information obtained from social networking websites, including Facebook, when deciding whether and how to underwrite applicants for life insurance. For instance, an underwriter might notice on a social network that a certain applicant has indicated he or she is a fan of risky hobbies like skydiving or rock-climbing, or has a circle of friends who are active in these sports. This information might be useful to the underwriter in case the life insurance application indicates that the applicant does not have risky hobbies.

Another disruptive change in the revenue stream is virtual currencies. LeHong, Research Vice President and Gartner Fellow, also pointed to the cryptocurrency Bitcoin (Nakamoto, 2008) as a major disruptive force. According to Professor Campell Harvey of Duke University (Card, 2014), Bitcoin, like all cryptocurrencies, bears fewer risks than debit or credit card fraud and minimal cost of transacting compared to traditional methods adopted by traditional institutions. Cryptocurrency relies on the power of computers to make possible real time exchange of ownership, verification of ownership, as well as the ability to design algorithmically conditional contracts (Card, 2014). In addition, cryptocurrency-based transactions require triangulation of data to ensure security, accuracy and trust without using centralized institutions. All of the above issues will sire the number crunching and user profiling power of open data analytics, which ¯have a big role to play in analyzing the huge amounts of data that Bitcoin and cryptocurrency systems will be generating. Insurance companies will be able to issue and guarantee their own currencies, which they can spend with associated vendors.

Based on the analysis of a number of companies, Deloitte identifies different suppliers to use data to generate revenue, grouped around five archetypes (Deloitte, 2012):

- Vendors that publish data, including the public sector;
- Aggregators that pool publicly available data and combine them to produce useful insights to be used by the different users;

[9] http://www.lifehealthpro.com/2010/04/29/social-networking-a-life-underwriting-tool-report. Accessed 16 April 2016.

- Apps developers that enable users to make more informed decisions (for instance, apps building on crime data, transport data);
- Enrichers that are large and established businesses producing significant quantities of open data and combine them with their own proprietary sources to provide services (for instance, insurance companies, retailers); and
- Enablers that are organizations that do not monetize open data directly but provide platforms and technologies that others can use (for instance, websites that enable data sources of all types to make subsets of their data available to seek solutions also for insurance companies).

4.1.6. *Key resources*

Within the insurance companies, the sectors in which the analysis of open data is more common are CRM Analytics, Finance and Accounting, Vertical solutions for business and e-commerce. The challenges in many insurance companies pass through the growth of their resources. The innovation discussed in this chapter might require an infusion of new blood in the resources of insurance companies. In this perspective, there are core competencies for the search, collection, analysis, interpretation and use of information. Data analysis skills can be a valuable support for the interpretation and organization of a large amount of data and for the detection of solutions able to use the information obtained consistently with the business objectives.

To start with, open data can revolutionize employment relationships as more and more companies can rely on crowdsourcing; thus, outsourcing a function once performed by employees to an undefined (and generally large) network of people in the form of an open call. This can take the form of peer-production when performing the job collaboratively (Fretty, 2014). As organizations rely more and more on crowdsourcing, they will require a strong coordination mechanism, capable of gathering and reconciling diverse information and views. After all, insurance companies need to be able to judge data quality, find ways to overcome territorial differences and apply to organizational goals (Fretty, 2014).

4.1.7. *Key processes and systems*

Open data poses big challenges and requires technology investments: to access, process and analyze massive amounts of data. It also demands companies to make changes on business processes to capitalize on open data. For example, it will require a process to scout the open data available and have an internal infrastructure to host the relevant data.

At the same time, it is necessary to also have strong tools for data analytics. They are essential to be able to extract useful information from diverse types of data and formats.

Some technical tools and applications useful for processing open data are as follows:

- SQL query;
- Hadoop;
- OLAP systems;
- Systems of statistical analysis such as R, SAS and SPSS;
- Scripting languages such as bash, PHP, PERL and Python;
- Platforms for data management and strengthening;
- PMML (Predictive Model Markup Language);
- Platforms and Big Data applications (for instance, Hadoop, MapReduce, Splunk and Cassandra);
- Platforms for machine learning (for instance, Apache Mahout);
- Business analytics.

Some of these software are open source (for instance, Hadoop). For using other software, it would be necessary to buy a license. In both cases, it is necessary to consider the processing and storage costs and, of course, the human resources necessary.

Particularly important is to have a flexible and powerful ETL system (Vassiliadis, 2012). An ETL allows extracting data in different formats and types, transforms them and loads them in internal archives.

4.1.8. *Key partnerships and collaborations*

Insurance companies need to consider possible ecosystems suitable for markets today. In this model, multiple players collaborate.

Collaboration, in the meaning of "working together", is necessary in modern times. It is necessary to work with different vendors, partners, outsourcers, fintechs and so on.

A business model innovation creates the basis for alliance or partnerships between different financial and non-financial institutions. Insurance companies will need to agree or extend, for instance, on partnerships with the entities providing open data and/or technology vendors that can supply and service connected devices, which can support other channels. They will also need to set up broader partnerships to secure direct access to customers and valuable information. The rising importance of ecosystems entails the risk that new players will enter the insurance industry at different points of the value chain. New players could also take control of these ecosystems — potentially leveraging far more detailed customer insights than the ones available to insurance companies. The long-term result could be lower returns for insurance companies if they lose control of the relationships with the customer. As mentioned before, open data analytics can help in this respect.

Further developments in advanced predictive open data analytics are likely to lead to complex decision-making scenarios based on a variety of stress tests in multiple market conditions against several key performance indicators (KPIs). Such business simulations can support the design and redesign of insurance operations (Groves *et al.*, 2014). For example, by simulating different market and competitor scenarios, open data can assess business agility and resilience of given products or risks.

4.1.9. *Costs and risks*

As open data, information and knowledge become the focal point for collaboration at all levels, punting for intellectual property rights, data and intangible assets will likely become a focal point (Viscusi & Batini, 2014).

Whether the reference is to life or health, property (home, motor, marine, aviation) or casualty/liability insurance, the core issue is that the involved items (life of the person, health of the person, property, liability of the person/organization) can have a large set of potential outcomes. There is the need for compensation if negative outcomes occur in practice. The core issue then is estimating the probability distribution of the potential losses from the data submitted at the point of underwriting the insurance policy (and later during the term of the insurance policy if agreed at the point of underwriting). It is critical to use all this to drive the risk selection decisions, risk pricing decisions, reserving decisions, exposure management decisions, capital allocation decisions and so on.

Fraud management is another important field where open data can support. In some countries, it is required to report them to a central insurance regulator (such as IVASS in Italy). It is possible to use them to evaluate and fight potential costs from this aspect.

The integration of open data can lead to billions of data being available to be shared in a unique system in which the information analysis, which takes into account the specific nature of each country, may allow the emergence of predictive models, useful to "anticipate" the data (in claims) in order to increase profitability. A court record violation is created by a jurisdiction (a local county, city or state, for example) when a citation is given, such as when a driver receives a traffic citation from a police officer, and that citation is uploaded into the jurisdiction's case management system. Open data analysis can use court record violation data (both criminal and traffic), homeowners policy and performance data, and household matching logic. A third party, such as Trans Union, can provide this type of service to help predict homeowners' policy claims and losses.[10]

With these approaches, predictive modeling can take into account not only customer purchasing patterns and feedback, but also an increasing number of disruptions due to civil unrest, natural disasters or even sudden economic developments. Open data analytics can keep an eye to track the developments in any of the critical

[10] http://newsroom.transunion.com/new-transunion-analysis-finds-using-court-record-violation-data-helps-predict-homeowners-policy-claims-and-losses. Accessed 16 April 2016.

risk factors that can compromise the ongoing concern of the organization from a variety of sources (for instance, social media, blogs, weather forecasts, news sites, stock trackers and the like). It then can alert and instigate risk management scenarios in support of underwriting or fraud prevention.

5. Some Experiences in the Use of Open Data in Insurance

5.1. *The use of open data in retail insurance*

Reverberi & Russo (2015) report of the project Energie Sisma Emilia (www.energie.unimore.it). It analyses the data on contributions paid for the reconstruction of economic activities damaged by the earthquake that hit Emilia Romagna, an Italian region, in 2012. These are essential information for monitoring the reconstruction after the earthquake. It is interesting to consider the criteria to assign contributions to the restoration and reconstruction of the buildings (for industrial, agricultural and commercial) damaged by an earthquake. The data collected during the submission of applications for assistance, then processed by the procedure for its payment, may allow drawing a clear picture of the reconstruction process. They can be used to highlight the specific sector and size of businesses affected, in the municipalities of the crater of the earthquake as well as of construction companies and professionals involved in the process of submitting applications for assistance and the implementation of the intervention.

5.2. *The use of open data in wholesale insurance*

The digital revolution is spreading around the world. Insurance is no exception and there are recent studies that aim to create and describe a model for Digital Insurance (Nicoletti, 2016). The spread of digital insurance is following the path of the so-called Information and Communication Technologies (ICT) consumerization. Contrary to the past in which ICT innovations were first used in the businesses, more and more, customers get priority in introducing ICT innovation. This has happened with physical products, like the smartphones and the tablets. Initially, consumers were the main target. Later, they expanded also in the use of businesses. Something similar is happening in financial services, like in banking, where Retail Banking was the first user. Only now, solutions are introduced also for corporate, small and medium-size firms.

Open data can be used also to support the wholesale insurance services. Some start-ups are already moving in this direction. More will come. Wholesale financial services are essentially operators which provide coverage for risks outside of the preference of admitted companies or which provide specialized capabilities.

In order to analyze how digital insurance could support this sector, it might be interesting to use the same approach introduced before based on Rudyard Kipling's five *W*s and one *H*. In the case of Digital Wholesale Insurance, this would mean to answer the following questions:

- Why: The reason to use open data could be defensive for wholesale insurance companies — start-ups might be able to invade their markets and introduce disruptive innovations. As a matter of fact, the use of open data could help also incumbent players in wholesale insurance to become more effective, efficient and economical;
- Where: Globalization, thanks to cloud computing, can easily expand to the reach of concentrated markets, fueling competition but also increasing the size of markets;
- What: data analytics also using open data can help improve the analysis of the risks, data analytics can be applied to capital markets, security, customer insights, channel marketing, and providing new datasets for risk pricing and tracking. Open data can be a basis to encourage insurance companies to leverage their sophisticated underwriting capabilities to understand and insure against more complex risks;
- Who: Open data available in social media, now spreading in financial services, could help in the case of wholesale insurance. An open data "marketplace" could help also in the wholesale

insurance market to find the best solution for the requirements of a specific customer;
- When: Open data can help in reducing the time to decision and especially to take decisions when the operator wants;
- How: Making use of open data could change in a radical way and also the wholesale insurance environment.

Open data are not only a way to improve the processes of the wholesale insurance companies. They can support the introduction and improvement of new products, for instance, insurance for cyber risk by supporting the analysis of several data available on the web on the subject.

Of course, in this new open data environment, new business models would emerge. They could foster, for instance, a much closer collaboration and partnership between wholesale insurance companies and brokers. The latter could greatly benefit from digital technologies but brokers are not always able to invest in them, due to the very small size or to the lack of the necessary knowledge.

An open data approach can help in the UNEP FI Principles for Sustainable Insurance (PSI).[11] This is a global framework for the insurance industry to address environmental, social and governance risks and opportunities. With PSI, it is possible to develop new products (Bates, 2012). The Lloyds of London, along with others, are lobbying to have their position in the market for meteorological data improved in order that they would be able to develop insurance products around extreme weather events (BERR, 2008) — a potentially lucrative sector for investment as climate change is an important issue.

6. The Data Scientists

There is a growing awareness in the market about the need to invest in the analysis of data. Often there is a lack of specific skills and

[11] http://www.unepfi.org/psi/vision-purpose/. Accessed 16 April 2016.

governance models. In fact, according to the Observatory research Big Data Analytics & Business Intelligence (2014), only 17% of companies have adopted a Chief Data Officer and only 13% a Data Scientist (Accenture, 2014).

When companies decide to exploit open data, it is necessary to deal with the change of skills in the type of resources used to support this job. Analyzing information according to the "traditional" mode means to work on a given structured mode. This does not meet the complexity related to open data. The key job in open data is the so-called Data Scientist. Data Scientists are the people who understand how to find out answers to important business questions from today's large amount of structured and unstructured information (Davenport & Patil, 2012). As companies rush to capitalize on the potential of open data, the largest constraints they may face is the scarcity of these special talents.

The Data Scientists are change agents. The disciplines of change management and project management will continue to merge and consequently expand the role of the Data Scientist. Projects are inherently about change. A good Data Scientist has strong business acumen. They should be very capable of challenging a project deliverable. The true objective of the Data Scientist is to deliver business value.

It is interesting to consider the profile of Data Scientists as described in the WSP-G3-024.[12] A Data Scientist is a professional position that owns the collection, analysis, processing, interpretation, dissemination and display of organization's quantitative or quantifiable data for the analytical, predictive or strategic purpose. In the case of open data, the Data Scientist is essential since the data are external to the insurance company and need to be "discovered" as it is done in the scientific world.

Process modeling is a key skill used by the Data Scientist to ensure that the organization is able to meet business needs effectively. Through continuous process improvement, benchmarking and process modeling, the Data Scientist ensures that all tasks and activities

[12] http://www.skillprofiles.eu/stable/g3/en/profiles/WSP-G3-024.pdf. Accessed 14 April 2016.

carried out by the business address the company's needs. The Data Scientist must be knowledgeable and skilled in modeling the business area to identify business needs, problem areas, business requirements, opportunities for improvement and solution assessment.

The mission of the Data Scientist apart from the mentioned activities is also the development of predictive models to generate organized systems of advanced knowledge. With an excellent understanding of the business and/or mission of the organization, the Data Scientist should identify and access data sources able to support and develop a particular business process. He or she chooses methods and models most suitable and effective for guiding the company's strategic decisions, develop lines of evolution and operational plans, abstract the information obtained and, through these, generates directions and development programs of action.

Predictive analytics is the future. Data Scientists and enterprise architectures (EAs) will need to become well versed in the art and science of predictive analytics if their insurance companies wish to stay ahead of the game. The field of predictive analytics has been around for a long time, but the discipline is now exploding, as data analysis becomes a core function in virtually every organization. Predictive analytics is not just about data mining and using it to predict future events. It covers areas such as descriptive modeling and decision modeling. New technologies are coming online every day. A Data Scientist needs to be well updated in them and build core competencies in these disciplines.

The Data Scientist must have several capabilities:

- They must have business sense and understand what is creating value for the customer and for the insurance company.
- They must be able to elicit and validate the requirements needed to support solution development and document them in the format most conducive to solution development.
- The ability to evaluate risk is critical to a Data Scientist's success. The Data Scientist should provide risk decision support in collaborating with the businesspersons. This requires knowledge of methods and techniques for identifying insurance risks, defining a

risk management approach; and performing risk requirements planning, monitoring and control. The Data Scientist may be involved in risk decision support. Data Scientists, for example, can support the strategic planning process with the environment and competitive analysis and benchmark studies. They may become involved in preparing the business case for future initiatives and monitoring business results after an initiative implementation.
- In terms of data, they must be able to plan and prepare the data analysts adequately to meet the needs of the insurance companies. They must be capable of identifying significant data and the best sources for them. They must then collect the data, validate, analyze them, interpret them and build quantitative models. Therefore, the Data Scientist must know statistical methods and decision-making methodologies.
- Communication is a key interpersonal and organizational skill needed for the Data Scientist's success as a facilitator, presenter and negotiator. They must be able to communicate in an appropriate form (graphical or textual) the results emerging from the analysis, the modeling and the impact of the solution to their stakeholders.
- The Data Scientist must be ICT well informed and good in team working for effectively collaborating with the technical members of the team: such as the business analyst, the database designers and testers.

In terms of reporting, the Data Scientist normally reports to:

- Head of the strategies/marketing/controller (in large companies), or to the
- Manager of the business/sales (in medium companies).

7. The Governance of the Open Data Initiatives

Open data governance is a part of a broader information governance program that formulates policy relating to the optimization, quality, security, compliance and monetization of open data by aligning the objectives of multiple functions involved, such as strategy, marketing, sales, operations and so on.

Organizations also need to create and follow appropriate policies and procedures to prevent the misuse of open data, considering regulatory and legal risks when handling social media, geolocation, biometric and other forms of personally identifiable information. Organizations should define the governance of power users of open data, such as the Data Scientists (Soares, 2013). An open data governance policy should obey the organization's legal and regulatory requirements. An open data governance policy might state that an organization will not integrate a customer's personally identifiable information into his or her master data record without the respect of the appropriate compliance. Organizations need to optimize and improve the quality of their open data in the following areas:

- Metadata: to build information about inventories of open data;
- Data quality management: to cleanse, whenever possible, normalize and finally manage open data;
- Information lifecycle management: to archive and sunset open data when it does not make any more sense to retain their massive volumes in internal files.

Open data must be profitable for the organization. It is important not to waste resources to use open data if they are not useful to add value to the customers and the insurance company.

8. The Benefits

The biggest challenge facing the insurance industry is how to operate in an increasingly digital world. Open data can provide the right digital support from which insurance companies and brokers can rapidly respond and adapt to a changing marketplace, and deliver a compelling customer experience.

ABI's Director General, Huw Evans, urged insurance companies to "maximize the potential of the data revolution".[13] The ubiquitous push from governments to release more data publicly has resulted in new UK government and industry bodies being set up to address the use of open data for both the citizens and commercial purposes.

The benefits of open data are essentially five, labeled as the 5 *V*s as follows:

- Value: Open data is data freely available. Even taking into account the costs of getting, analyzing and processing them, the total costs of operations are decidedly lower than the ones of other types of data;
- Variety: Open data are available in different and distant sectors to help in different aspects of the insurance business;
- Velocity: There is no comparison to the slowness of other forms for getting information on the environments with respect to a digital analysis of existing open data;
- Veracity: Every open data normally is based on some real life situation. So open data tends to be rather reliable;
- Visibility: The data can be shared publicly or selectively.

9. The Future

This chapter examines briefly some uses of open data to transform insurance digitally. Rapidly emerging technologies for open data are becoming enablers to creating competitive advantages. An easily accessible source of aggregated open data that provides insight opens up the imagination and limitless potential for the use of open data to succeed also in the insurance business. The future looks very exciting with new technologies that could change the ways that insurance companies work today.

[13] https://www.doorda.com/open-data-competitive-advantage-for-insurance companies/. Accessed 14 April 2016.

Looking into the future, there is a technology developed for a specific case, the Bitcoin virtual currency, which has the potential to change many fields including insurance. The blockchain technology is interesting in itself. It is a clear example of how data accessible to multiple parties can change the processes and even the industry. The blockchain is a digital online, distributed and encrypted public ledger. It can maintain a sequential list of transactions replicated frequently among the different nodes. It can certify the correctness of information thanks to its presence on various network nodes. This is possible due to the fact the data is not kept in a single register center, maybe private, but in its pure application, it is made public with copies in each node of the network. The fairness of a transaction is derived from the so-called consensus method: "if everyone knows it, it must be true".

Blockchain can be open data affecting the interactions between insurance companies and third parties: agency networks, external vendors and customers. It can affect the portfolio management, administration, sales and claims handling. In insurance, Blockchain could help in setting up smart contracts with a distributed ledger solution. It could help in managing customer identities, reference data and assets, increasing secure visibility, ensuring a seamless, reliable and uninterrupted messaging service to the insurance market. It is a very cost-effective method of facilitating the availability and exchange of data between many parties interested in the insurance business. It is a trusted utility service that boosts insurance market competitiveness.

The applications of the Blockchain, in all sectors of insurance could be many. Figure 3 shows the typical high-level process of insurance. Blockchain could be useful along the entire cycle.

Blockchain can be applied to the time of the customer application, to verify the identity of the customers or to limit the risks of fraud.

Blockchain can support Bancassurance underwriting. This is the case of Tradle. This start-up has closed the gap with where you want your personal information stored with its high resilience to potential hackers and cyberattacks (Grisoni, 2016). Allowing the

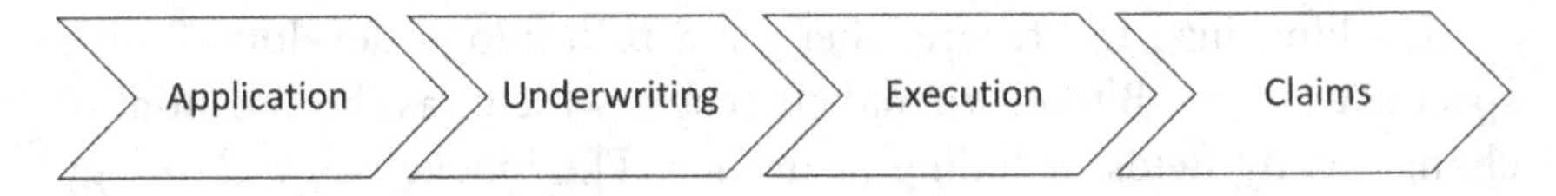

Fig. 3. The insurance process

partner banks to share data about their customers, such as the document's identity, the partner company of the bank can offer a very quick and streamlined process for subscribing to a product without asking data already available in possession of the information. Document exchange and communications could be saved and guaranteed by Blockchain, which certifies these mailings, allowing the recovery in full compliance with European data laws. Another example in underwriting is provided by SafeShare, a British company. It uses Bitcoin's underlying Blockchain technology to confirm counterparty obligations. Blockchain technology facilitates the delivery of a flexible and a responsive product to customers at a reasonable price.[14]

It can also be applied at the execution by the automation of the contracts for getting a reduction of administration costs for reconciliation and error. Smart contracts powered by a Blockchain could provide customers and insurance companies with the means to manage claims in a transparent, responsive and irrefutable manner. Contracts and claims could be recorded onto a Blockchain. The network could validate them, ensuring that insurance companies pay only valid claims (Deloitte, 2016).

Blockchain could be used in insurance for the automatic discovery of claims. Everledger, for example, uses the Blockchain to create a distributed ledger that records details of precious stones like diamonds (Davies, 2015). This ledger allows insurance companies (as well as potential purchasers) to check the history of any individual stone, including previous claims. In this way, it helps insurance companies prevent, detect and counter fraud.

[14] http://www.econotimes.com/SafeShare-Releases-First-Blockchain-Insurance-Solution-For-Sharing-Economy-181326. Accessed 15 April 2016.

Blockchain could be used together with connected devices: the car's black box, a wearable or an installed device in the house sensor to detect an anomaly and sends an alarm. A Blockchain would end up in the workflow associated with the complaint, or pre-complaint, automatic problem detected by the connected device.

Another example of a possible application of Blockchain technology is a "peer-to-peer" insurance: a business model in which groups of individuals make sure with each other by sharing the premium, similar to what Friendsurance does.[15] A portion of the premiums paid goes to the company for the coverage of greater gravity claims, the remaining part remains in a fund of mutuality to take care of lower claims. The commitment to contribute to the fund of the mutuality is signed and kept inside the Blockchain: the award is not paid immediately but "preserved" in the Blockchain by transferring the contributions from each individual at the bottom of the mutuality only in the event of a claim.

Insurance companies are laggard with respect to the banking world in the examination of the potential use of Blockchains. There are exceptions. Allianz stands out in France, with a start-up in this area. Lloyds is considering the redesign of their operating model with Blockchain. Probably a real impact of this technology on insurance will take at least 5 or 10 years, even if people are now accustomed to shorter and shorter cycles for the introduction of new technologies.

10. Conclusions

This chapter is a contribution to the continuing debates and analyses on how to use open data in insurance companies (Cole, 2012). Open data might change in several geographies and situation, but their availability will increase.

---

[15] http://www.celent.com/system/files/back_to_the_future.pdf. Accessed 15 April 2016.

In conclusion, actual practices surrounding open data can produce a wide-ranging diversity. There will be an increase over time of the availability of open data. The insurance companies should increase their ability to gather and make use of it through advances in resources, human and non-human, such as Data Scientists, software, scientific understanding and the like. Given the current world situation, the concerns about its misuse, its errors and the unevenness of resources to exploit it will not decrease. Therefore, there will be varying degrees of openness to public data in the near future.

The Big Data Analytics, and in particular open data, undoubtedly represents a major challenge for insurance companies, with important implications from several points of view. They are also an extremely important opportunity. In times of rapidly changing market dynamics, an opportunity is welcome. Making use of open data allows exploiting a treasure of data helping in getting a real competitive advantage. As noted by Greg Baxter, Citi's Global Head of Digital Strategy, we are not even at "the end of the beginning" of a digital disruption.[16]

References

Accenture (2014). Big data analytics in supply chain: Hype or here to stay? *Accenture Global Operations Megatrends Study.*

Bates, J. (2012). "This is what modern deregulation looks like": Co-optation and contestation in the shaping of the UK's Open Government Data Initiative. *The Journal of Community Informatics*, *8*(2).

Belissent, J. (2013). Open data is not just for governments anymore. Retrieved from http://blogs.forrester.com/jennifer_belissent_phd/13-02-21-open_data_is_not_just_for_governments_anymore. Accessed 12 April 2016.

Belissent, J. (2016). Retrieved from http://blogs.forrester.com/jennifer_belissent_phd/13-02-21-open_data_is_not_just_for_governments_anymore. Accessed 2 April 2016.

[16] http://www.agefi.fr/sites/agefi.fr/files/fichiers/2016/03/citi.pdf, Accessed 16 April 2016.

BERR (2008). Supporting innovation in services. *Department for Business, Enterprise and Regulatory Reform*, London, UK.

Card, J. (2014). Bitcoin: a beginner's guide for entrepreneurs. https://faculty.fuqua.duke.edu/~charvey/Media/2014/Guardian_October_17_2014.pdf, Accessed 11 February 2019.

Chandler, A. D. (1990). *Strategy and Structure: Chapters in the History of the Industrial Enterprise*, Vol. 120. Cambridge, MA: MIT Press.

Christensen, C. (2013). *The Innovator's Dilemma: When New Technologies Cause Great Firms to Fail.* Cambridge, MA: Harvard Business Review Press.

Cole, R. J. (2012). Some observations on the practice of "Open Data" as opposed to its promise. *The Journal of Community Informatics*, *8*(2).

Cummins, S., Peltier, J. W. & Dixon, A. (2016). Omni-channel research framework in the context of personal selling and sales management: A review and research extensions. *Journal of Research in Interactive Marketing*, *10*(1), 2–16.

Davenport, T. H. & Patil, D. J. (2012). Data Scientist. *Harvard Business Review*, *90*, 70–76.

Davies, S. (2015). Bitcoin: Possible bane of the diamond thief, *Financial Times*. Retrieved from http://www.ft.com/cms/s/0/f2b0b2ee-9012-11e4-a0e5-00144feabdc0.html#axzz3Qm7XPPbZ. Accessed 14 April 2016.

Deloitte (2012). Open growth. Stimulating demand for Open Data in the UK. *A briefing from Deloitte Analytics.*

Deloitte (2016). Blockchain applications in insurance. Retrieved from file:///C:/D/My%20Documents/Businesses/Financial%20Services/Fintech/Book/Innovation/Processes/Blockchains/ch-en-innovation-deloitte-blockchain-app-in-insurance.pdf. Accessed 16 April 2016.

Fretty, P. (2014). Merging crowdsourcing with big data analytics. Retrieved from http://www.iee.ch/etech/2014/etech_0314/ind-1.htm. Accessed 16 April 2016.

Grisoni, A. (2016). Se anche le assicurazioni guardano alla blockchain. *AziendaBanca*, March, 66:67.

Groves *et al.* (2013). The "Big Data" Revolution in Healthcre – Accelerating Value and Innovation. McKinsey & Company, Chicago, IL.

Gurstein, M. B. (2011). Open Data: Empowering the empowered or effective data use for everyone? *First Monday*, *16*(2).

Kipling, R. (2013). *Just So Stories*. CreateSpace Independent Publishing Platform.

Leavitt, H. J. (1965). Applied organizational change in industry: Structural, technical and humanistic approaches. In *Handbook of Organizations*, ed. March, J.G., Routledge, Sage, Los Angeles, CA: London, UK, 1144–1170.

McDowell, C. (2015). Where there is Open Data, there is Competitive Advantage for Insurance companies, 23 October. https://medium.com/doorda/where-theres-open-data-there-s-competitive-advantage-e47f26ced55b. Accessed 11 February 2019.

Morabito, V. (2015). *Big Data and Analytics: Strategic and Organizational Impacts*. Cham, Switzerland: Springer.

Nakamoto, S. (2008). Re: Bitcoin P2P e-cash paper. *Email posted to listserv*, *9*.

Nicoletti, B. (2014). *Mobile Banking*. London, UK: Palgrave MacMillan.

Nicoletti, B. (2016). *Digital Insurance*. London, UK: Palgrave MacMillan.

Osterwalder, A., Pigneur, Y. & Clark, T. (2010). *Business Model Generation: A Handbook for Visionaries, Game Changers, and Challengers*, Hoboken, NJ: Wiley.

Papastefanatos, G., Vassiliadis, P., Simitsis, A., & Vassiliou, Y. (2012). Metrics for the prediction of evolution impact in etl ecosystems: A case study. *Journal on Data Semantics*, *1*(2), 75–97.

Reverberi, M. & Russo, M. (2015). I contributi alle imprese colpite dal sisma del 2012 in Emilia-Romagna: una base informativa per l'analisi e il monitoraggio della ricostruzione (No. 0069). University of Modena and Reggio Emilia, Marco Biagi Department of Economics.

Soares, S. (2013). *IBM InfoSphere: A Platform for Big Data Governance and Process Data Governance*. Boise, ID: MC Press Online, LLC.

Ubaldi, B. (2013). Open government data: towards empirical analysis of open government data initiatives, OECD Working Papers on Public Governance, No. 22, OECD Publishing.

Viscusi, G., & Batini, C. (2014). Digital information asset evaluation: characteristics and dimensions. In *Smart Organizations and Smart Artifacts* (pp. 77–86). Springer, Cham.

Web Sites

Last accessed 15 April 2016
www.cloudmate.com
www.data.com
www.leandigitize.com
www.skillprofiles.eu
Data.Gov (2011). Retrieved from http://www.data.gov/. Accessed 12 April 2016.

Open Parliament.ca. Retrieved from http://openparliament.ca/. Accessed 16 January 2016.

Profile Sheet Wsp-G3-024 Data Scientist, http://www.skillprofiles.eu/stable/g3/en/profiles/WSP-G3-024.pdf. Accessed 14 April 2016.

The Data.Gov. (2011). UK Web site Retrieved from http://data.gov.uk/. Accessed 12 April 2016.

Chapter 5

Open Public Sector Information in Australia*

John Gilchrist

Australian Catholic University, Australia
Barrister and Solicitor of the
High Court of Australia
and of the Supreme Court of the Australian Capital Territory, Australia

The most significant changes effecting public access to public sector information and data in Australia during this century have not been made through copyright law reform but through changes in copyright practice. The changes brought about by licensing reflect the changes brought world-wide by the open content movement. It is important for the future of Australian reform that copyright law and policy are consistent with, and not a barrier to, access to public sector information and data if the goals of accountability of government, better communication and interaction with the community and net benefits to the economy and society are to be promoted.

*Readers should be advised that the law and developments described in this Chapter were those available as at 27 October 2016.

1. Copyright in Australian Government Information and Data

Like some other former British territories, Australia developed a copyright system based on an English model with the concept of "fair dealing" defenses to infringement. Protection of copyright works and other subject matter has historically been of a high level; the British colonies of Australia were bound by the Berne Convention as an emanation of the Britain's adherence to that Convention in the late 19th century and subsequently Australia was bound as an independent country.

In the 19th century, this English copyright model protected government works under the general provisions of the law in a similar way to other legal persons, and on the passage of the Australian federal *Copyright Act 1912*,[1] which adopted the British *Copyright Act 1911*, special provisions were made for government ownership of copyright. These provisions protected works in a similar way to works produced by the private corporate sector. Under the current *Copyright Act 1968*, this regime was broadened to include a wider subject matter. Government literary works as well as government artistic, dramatic and musical works are protected for 50 years from publication (or in the case of artistic works from the making) and other subject matter such as films and sound recordings are protected for terms similar to non-government subject matter. Government information — sometimes referred to as public sector information (PSI) — mostly falls under the definition of literary works under the *Copyright Act 1968* (which includes computer software) but it may exist in other media forms such as in sound recordings. Data accumulated in a computer readable form falls within the definition of "literary work" under

[1]Under s51(xviii) of the Australian Constitution 1900 (UK), the Australian federal parliament has power to make laws with respect to copyrights, patents of inventions and designs, and trademarks.

that Act. In broad terms, apart from limited library and archive copying, more than insubstantial use by the public under copyright law to public sector information has laid within the fair dealing defenses which are limited to research, study, criticism, review, parody or satire and reporting of news, and the boundaries of the research and study defense has been partially clarified under amendments to the law. That and other defenses are reliant on case law clarification. The fair dealing defenses under the Australian law are narrower in application than the "fair use" doctrine in the United States under its Copyright Act of 1976. A useful comparison of the two is set out in the draft Productivity Commission Report on Intellectual Property Arrangements of April 2016 (Productivity Commission, 2016). In the 20th century, the unremunerated licensing of public sector information was largely focused on legislative, executive and administrative materials to law publishers until the early 1980s when limited open licenses were given by the federal and some state governments to publishers and educational establishments to reproduce these materials. Government licensing of the use of other government information by the public was normally done on an individual request basis, usually subject to conditions and often remuneration. In the late 20th century, this was done on a centralized basis through the Department of Communications, Information Technology and the Arts (Commonwealth Copyright Administration (CCA)).

The most significant changes affecting public access to public sector information and data in this century have not been through the copyright law, but through copyright practice. The licensing of copyright information by state and federal governments over the last decade in this century has far surpassed changes in the law. The changes brought about by licensing reflect changes brought worldwide by the open content movement and in particular by the now widespread use of creative commons licensing.

2. The Development of Australian Open Public Sector Information

The steps leading to these changes have occurred over more than a decade[2]:

- In 2004, Stage 1 of a Project known as Government Information Licensing Project (GILF) was initiated by the Queensland Spatial Information Council to review licensing practices and options in its business environment. It found inconsistent licensing practices by Queensland Government agencies. In 2005–2006, Stage 2 of the GILF Project resulted in a recommendation that state government agencies pilot the move to an information licensing framework based on Creative Commons for qualifying information where no issues of privacy, confidentiality or other legal or policy constraints applied.
- In April 2005, the Australian government released the report of the Copyright Law Review Committee on its *Crown Copyright* reference. The Committee recommended the abolition of the specific government ownership and subsistence provisions in the *Copyright Act 1968*, the abolition of government copyright in certain judicial, executive and legislative materials and changes in the management practices of state, territory and federal governments dealing with government copyright material.
- In September 2008, a review of the National Innovation System, entitled *Venturous Australia: Building Strength in Innovation* (the "Cutler Report"), stated "Australia is behind many other advanced countries in establishing institutional frameworks to maximize the flow of government generated information and

[2]For an expanded summary of these steps to 2014, refer to John S Gilchrist, *The Government and Copyright* (Sydney University Press, 2015) xxi–xxviii. For a detailed review of the literature in Australia and selected jurisdictions on open access policies, see Anne Fitzgerald, *Open Access Policies, Practices and Licensing*, (Cooperative Research Centre for Spatial Information, QUT, 2009) and the two volume collection Brian Fitzgerald (ed.), *Access to Public Sector Information: Law, Technology and Policy*, (Sydney University Press, 2010).

content".[3] Its recommendations included that Australian governments should adopt international standards of open publishing as far as possible and material released for public information by Australian governments should be released under a creative commons license.[4] In its response of May 2009 in *Powering Ideas: An Innovation Agenda for the 21st Century,* the Australian government stated that it controlled "mountains of information, and it is determined to make more of this vast national resource accessible to citizens, business people, researchers and policy makers" and announced it would take steps to develop a more coordinated approach to Commonwealth information management, innovation and engagement involving the Australian Government Information Management Office (AGIMO) and other federal agencies (Department of Innovation, Industry, Science and Research, 2009).

- The Economic Development and Infrastructure Committee (EDIC) of the Victorian parliament in its report of June 2009, entitled *Inquiry into Improving Access to Victorian Public Sector Information and Data,* stated:

---

[3]Australia. Cutler & Company Pty Ltd, *Venturous Australia: Building Strength in Innovation* (2008), Report to Senator the Hon Kim Carr Minister for Innovation, Industry, Science and Research (the Cutler Report): 94 http://apo.org.au/research/venturous-australia-building-strength-innovation or www.innovation.gov.au/Innovation/Policy/Pages/ReviewoftheNationalInnovationSystem.aspx.

[4]The Cutler Report followed earlier reports which pointed out the advantages to be gained from reuse of government-held materials in the digital content sector such as *Commerce in Content: Building Australia's International Future in Interactive Multimedia Markets,* Cutler & Company Pty Ltd, Melbourne, 1994, which recommended that government provide access to culturally significant data in digital form to IMM content developers and users by early digitalisation of national collections and archives (at 43) and Department of Communications, Information Technology and the Arts, *Unlocking the Potential: Digital Content Industry Action Agenda, Strategic Industry Leaders Group report to the Australian Government* (2005) 46–47, http://www.archive.dcita.gov.au/2007/12/unlocking_the_potential_digital_content_industry_action_agenda_report where it reported that there were insufficiently developed mechanisms for accessing Crown IP for exploitation by digital content firms and proposed work in the area of alternative approaches to intellectual property licensing, such as the Creative Commons.

The Committee believes that open access should be the default position because:

- PSI is publicly funded and is generated for the purpose of administering the state and undertaking core functions of governance. As a resource created on behalf of all citizens, PSI should be accessible to all citizens; and
- Economic and social benefits arising from the release of the Victorian Government PSI will likely outweigh the benefits of treating it as a commodity (Economic Development and Infrastructure Committee, 2009).

EDIC also recommended a consistent copyright licensing system over government information for use across all government departments, developed and administered through a central office (Economic Development and Infrastructure Committee, 2009, p. xxvi).[5]

In February 2010, the Victorian government tabled its response which agreed that the default position for the management of PSI should be open access. The Victorian government committed itself to the development of a whole-of-government Information Management Framework (IMF), whereby public sector information is made available under Creative Commons licensing by default with a tailored suite of licenses for restricted materials (Department of Innovation, Industry and Regional Development, 2010).[6]

---

[5]**Recommendation 11:** That the Victorian Government develop a consistent copyright licensing system for use across all government departments.
Recommendation 12: That the Victorian Government establish a central office to develop a copyright licensing system, and provide advice to government on government copyright.

[6]Victoria. Department of Innovation, Industry and Regional Development, Parliament of Victoria, *Whole of Victorian Government Response to the Final Report of the Economic Development and Infrastructure Committee's Inquiry into Improving Access to Victorian Public Sector Information and Data* (2 February 2010) 8 http://www.parliament.vic.gov.au/edic/inquiries/article/1019.

- In July 2007, the Council of Australian Governments, Online and Communication Council commissioned the development of a national information-sharing strategy. This was aimed at promoting better government service delivery and improved policy development through focused interagency collaboration and was widely supported across agencies and jurisdictions. In August 2009, AGIMO published a report, entitled *National Government Information Sharing Strategy*, endorsing nine information-sharing principles aimed at providing benefits to governments and the public. Included among the principles were the following: (1) agencies should facilitate whole-of-government approaches to information management through interdepartmental communication, collaboration and consistency across government; (2) agencies should promote information reuse, that is, agencies need to investigate the conditions of use they should apply to the different elements of their information catalogue, for example, legislation, classification, freedom of information and licensing requirements; and (3) agencies should do so ensuring privacy and security requirements are met (Department of Finance and Deregulation, 2009b).
- In December 2009, the Australian (federal) government's 2.0 taskforce delivered its final report, entitled *Engage: Getting on with Government 2.0*, whose central recommendation was a declaration of open government by the Australian government stating that:
  - Using technology to increase citizen engagement and collaboration in making policy and providing service will help achieve a more consultative, participatory and transparent government;
  - Public sector information is a national resource and that releasing as much of it on as permissive terms as possible will maximize its economic and social value to Australians and reinforce its contribution to a healthy democracy;
  - Online engagement by public servants, involving robust professional discussion as part of their duties or as private citizens, benefits their agencies, their professional development, those with whom they are engaged and the Australian public.

This engagement should be enabled and encouraged (Department of Finance and Deregulation, 2009a).

The report noted that meeting these key points at all levels of government was integral to achieving the Government's objectives, including public sector reform, innovation and using the national investment in broadband "to achieve an informed, connected and democratic community" (Department of Finance and Deregulation, 2009a).

- In July 2010, the then federal Minister for Finance and Deregulation released a *Declaration of Open Government,* which implemented this recommendation. Later in the same year, the *Statement of Intellectual Property Principles for Australian Government Agencies* (Attorney General's Department, 2010)[7] was amended to reflect government decisions on the free use of public sector information. The *Statement* advised agencies to license public sector information under Creative Commons BY license (otherwise known as the "Attribution Licence") or other open content licenses (Attorney General's Department, 2010) and also stated that when federal (Commonwealth) records become available for public access under the *Archives Act 1983*, public sector information covered by government copyright should be automatically licensed under an appropriate open content license. In January 2011, the federal Attorney General's Department released *Guidelines on Licensing Public Sector Information for Australian Government Agencies* to assist agencies in implementing this policy (Attorney General's Department, 2012).
- On 29 June 2012, the then federal Attorney General gave a reference to the Australian Law Reform Commission (ALRC) for inquiry and report into *Copyright and the Digital Economy*. The

[7]The *Statement* provides a policy for the management of intellectual property across Commonwealth agencies and particularly addresses the contracting practices of the Commonwealth. The *Statement* was amended on 1 October 2010 to reflect Government decisions in relation to ownership of IP in software procured under ICT contracts (principle 8(a)) and free use of public sector information (principle 11(b)).

Terms of Reference required the ALRC to report by 30 November 2013 on "whether the exceptions and statutory licenses in the *Copyright Act 1968* are adequate and appropriate in the digital environment", having regard among other things to "the importance of the digital economy and the opportunities for innovation leading to national economic and cultural development created by the emergence of new digital technologies" (Australian Law Reform Commission, 2013, p. 5, 6). In its final report, the ALRC made a number of recommendations for the repeal of a range of specific exceptions and the modification of various statutory licenses in the *Copyright Act 1968*. They recommended a broad "fair use" exception, covering in a non-exhaustive list, research or study, criticism or review, parody or satire, quotation, reporting news, professional advice, non-commercial private use, incidental or technical use, library or archive use, education, and access for people with disability. Under these proposals, whether a use is fair will be determined by factors similar to those presently set out in Section 107 of the *Copyright Act of 1976* (US), that is, the purpose and character of the use, nature of the material used, the amount and substantiality that is used and the impact on any potential market for the material (Australian Law Reform Commission, 2013, pp. 90–96). Copyright owner interests have objected to this proposed relaxation in the law.

- These Australian developments occurred against a background of significant international reports and initiatives aimed at promoting better access to public information.[8] The reports and

[8]For a summary of these reports and initiatives to 2014, refer to John S. Gilchrist, *The Government and Copyright* (Sydney University Press, 2015) xxv–xxviii. See for example, the 2006 review by the UK Office of Fair Trading, *The Commercial Use of Public Information (CUPI)* (2006) Cm 4300, http://www.oft.gov.uk/shared_oft/reports/consumer_protection/oft861.pdf and the 2007 review, United Kingdom. Cabinet Office, *The Power of Information: An independent review by Ed Mayo and Tom Steinberg* (2007) http://www.epractice.eu/files/media/media1300.pdf and http://www.opsi.gov.uk/advice/poi/index.htm, the OECD, Directorate for Science, Technology and Industry, *OECD Recommendation of the Council for Enhanced Access and More Effective Use of Public Sector Information* (2008) and

initiatives have originated from a common understanding of the benefits of accessing and reusing environmental, spatial, technological and other scientific information produced by publicly-funded institutions, particularly in dealing with common and often global problems. They widened and gathered momentum in the light of the fundamental technological changes in the way people communicate and access information and are enabled to interact with government.

As a result of these developments, the federal (Commonwealth) and various state governments of Australia have adopted open content licensing policies for much of their government information using creative commons license guidelines. The federal government has devolved licensing of public sector information to individual departments and agencies and ceased the functions of the central clearing house model (CCA) for licensing.[9] Instead the *Guidelines on Licensing Public Sector Information for Australian Government Entities*[10] implement the default model of a free Creative Commons BY license.

The following *Guidelines* list examples of PSI for the purposes of the government's policy:

- Text-based publications: Government created reports, policy papers, budget papers, government-produced books providing

http://www.oecd.org/internet/ieconomy/40826024.pdf and Directive 2003/98/EC of the European Parliament and of the Council on the reuse of public sector information, of 17 November 2003.

[9]The CCA clearing house ceased to function in January 2011.

[10]In September 2016, the federal government issued *Guidelines on Licensing Public Sector Information for Australian Government Entities* (https://www.communications.gov.au/policy/policy-listing/australian-government-intellectual-property-rules) to assist government entities in implementing this policy. The term "entities" in the *Guidelines* is broadly defined as "Non-corporate entities subject to the Public Governance, Performance and Accountability Act (PGPA) are expected to comply with these guidelines. Corporate entities, previously subject to the Commonwealth Authorities and Companies Act 1997, may also consider these guidelines as an expression of good practice".

government information, text-based information on Government websites, Hansard, Explanatory Memoranda, Parliamentary reports, Official records of Parliamentary debates, Template forms;

- Legislation and legislative instruments;
- Forms of data: Spatial data, statistics, data products, databases, maps, administrative data;
- Audio–visual material that contains government information: Compact discs (CDs), DVDs, Videos;
- Visual material that constitutes government information: Photographs included in a government publication;
- Certain materials collected, preserved, and/or held by cultural institutions: Archival material that constitutes Government information;

Examples of material unlikely to constitute PSI for the purpose of the government's policy:

- Material held by cultural institutions for the value of its expression (e.g., a novel subject to third-party copyright ownership, an artistic work such as a painting or sculpture that was created pursuant to government funding or that has been preserved by a public institution);
- Software (Creative Commons licenses do not contain terms relating to the use of source or object codes. There are other open content licenses specifically designed for software that may be considered if the Commonwealth owns copyright in the software and the author agency determines that providing open access to the underlying code is appropriate. See, for example, the licenses available through the Free Software Foundation webpage, http://www.fsf.org/ and the Open Source Initiative webpage, http://www.opensource.org/);
- Artistic works that are Commonwealth copyright: Paintings and drawings, sculptures;
- Photographs exhibited as a work of artistic expression;
- Confidential material;
- Material that has national security and strategic interest implications;

- o Material that contains personal information;
- o Material that is commercially sensitive;
- o Material that is culturally sensitive.

Other relevant reasons for not treating material or data as PSI may include the incompleteness of material or data, such that it may be materially misleading.

Forms of data are expressly listed in the *Guidelines*. To encourage access to and reuse of public data, a federal government data portal was established presenting material under a Creative Commons Attribution Australia license.[11]

In December 2015, the federal government released a public data policy statement. This constituted a part of the national innovation and science agenda which the LNP conservative coalition embarked on and took to the 2016 federal elections. Public data sharing was perceived in the policy to promote innovation and better targeting of social services. "Publishing, linking and sharing public data can help make government more citizen-focused, create new and innovative products and services, and increase efficiency".

The public data policy included the statement:

> The Australian Government commits to optimize the use and reuse of public data, to release non-sensitive data as open by default, and to collaborate with the private and research sectors to extend the value of public data for the benefit of the Australian public.
>
> Public data includes all data collected by government entities for any purposes including government administration, research or service delivery. Non-sensitive data is anonymized data that does not identify an individual or breach privacy or security requirements.

[11]Following the *Declaration of Open Government* and as a response to the *Government 2.0 Taskforce Report*, refer Australia. Department of Prime Minister and Cabinet, data gov.au https://www.data.gov.au/about.

The public data policy requires Australian (federal) Government entities to

- make non-sensitive data *open by default* to contribute to greater innovation and productivity improvements across all sectors of the Australian economy;
- where possible, make data available with free, easy to use, high-quality and reliable Application Programming Interfaces (*APIs*);
- make *high-value* data available for use by the public, industry and academia, in a manner that is enduring and frequently updated using high-quality standards;
- where possible, ensure non-sensitive *publicly funded research* data is made open for use and reuse;
- only charge for *specialized data services* and, where possible, publish the resulting data open by default;
- *build partnerships* with the public, private and research sectors to build collective expertise and to find new ways to leverage public data for social and economic benefit;
- *securely share data* between Australian government entities to improve efficiencies, and inform policy development and decision-making;
- *engage openly with the states and territories* to share and integrate data to inform matters of importance to each jurisdiction and at the national level;
- *uphold the highest standards of security and privacy* for the individual, national security and commercial confidentiality; and
- ensure all *new systems* support discoverability, interoperability, data and information accessibility and cost-effective access to facilitate access to data.[12]

---

[12]At a minimum, Australian government entities will publish appropriately anonymized government data by default:

- on or linked through data.gov.au for discoverability and availability;
- in a machine-readable, spatially-enabled format;
- with high-quality, easy to use and freely available API access;

3. Recent International Open Data Reforms

The Australian federal government policy statement occurred in the context of recent international initiatives promoting accountability of government and access to government information and data. In May 2013, the then federal Attorney General announced that Australia would join the Open Government Partnership (Attorney General for Australia, 2013), a US Obama administration initiative in 2011 to promote accountability of government (to improve performance), transparency of government (to enable people to find and use information) and dispersal of knowledge among nations (The White House, 2011). From the original nations, there are now at least 70 members of the Open Government Partnership promoting open government reforms.

In 2013, the G8 nations developed an open data charter embodying five principles.[13] In 2015, this was augmented by an International Open Data Charter, a more inclusive document which allows for adoption by national as well as subnational governments.

- with descriptive metadata;
- using agreed open standards;
- kept up to date in an automated way; and
- under a Creative Commons By Attribution license unless a clear case is made to the Department of the Prime Minister and Cabinet for another open license.

Requests for access to public data can be made via data.gov.au or directly with the government entity that holds the data. If access to data is denied by an entity, users may appeal the decision using the public request functionality available through data.gov.au

[13]
1. Release open data by default: Government agencies should release all public data in open and machine-readable formats unless there is a compelling reason not to, such as national security or privacy concerns.
2. Ensure high quality and quantity of data: Government agencies should release a broad range of datasets that have been adequately vetted and cleaned.
3. Make data usable by all: Government agencies should standardize metadata (i.e., data that describes other data) for all datasets, use open licenses, and ensure general accessibility.
4. Release data for improved governance: Government agencies should share best practices on open data internationally, release certain "key datasets" specified in the charter, and seek input from civil society.

G20 nations, which include Australia, endorsed the six principles of the Open Government Data Charter in November 2015, as an important instrument in the fight against corruption.[14] As of October 2016, the international Open Data Charter has been adopted by 39 governments, 15 national and 24 local/subnational.[15] The principles of this and the G8 open data charters are consistent with the 2015 Australian federal policy.

The Australian policy reforms represent a significant change in access to government copyright material which stretches far beyond the limitations imposed by copyright law. They reflect a widespread change in governance values, from treating government information as a commodity to be sold, to an understanding of the benefits to government and to the community at large of the greater sharing of information held by government. While the government's policy provides that the default or starting position is that public sector information should be released free of charge under a Creative Commons "BY" license (the most liberal Creative Commons license)

---

5. Release data for innovation: Government agencies should release datasets considered "high value" by the charter, as well as engage with developer communities and fund open data start-ups.

All of the G8 countries had national data portals, with Canada as at 2014 having 214,033 datasets, the majority of which were geospatial datasets, such as map projections and coordinates. Daniel Castro and Travis Korte, *Open Data in the G8: A Review of Progress on the Open Data Charter* (March 2015) (Centre for Data Innovation, Washington), p. 3, 6. http://www2.datainnovation.org/2015-open-data-g8.pdf.

[14]Described as the "Open Data Anticorruption Package" ... "to strengthen access and use of open data in the anti-corruption, transparency and public integrity sector". Refer *Open Data Charter* http://opendatacharter.net/the-consultation-on-g20-anti-corruption-package-is-launched/ and *Introductory Note to the G20 Anti-Corruption Open Data Principles* www.g20.utoronto.ca/2015/G20-Anti-Corruption-Open-Data-Principles.pdf.

[15]*Open Data Charter* http://opendatacharter.net/adopted-by-countries-and-cities/. The size principles are: (1) Open by Default; (2) Timely and Comprehensive; (3) Accessible and Usable; (4) Comparable and Interoperable; (5) For Improved Governance and Citizen Engagement; and (6) For Inclusive Development and Innovation.

by its entities, it also states that entities should only apply Creative Commons "BY", or any other license, to a particular PSI following a process of due diligence and on a case by case basis. The Public Data Policy Statement aims to bring Australia in line with our international and jurisdictional counterparts.

As at 10 October 2016, the federal government data portal (www.data.gov.au) lists 268 organizations with 23,287 datasets, which comprise federal, state and local government entities. This web portal provides a central index to aid access by interested users. State governments and many semi-government entities have made similar initiatives with their own data websites.

In February 2016, the Australian Bureau of Communications Research published a critical review of studies on the economic impact of open government data — entitled "Open Government Data — Why it Matters". It concluded:

> While there is little consensus on the magnitude of the economic benefits of open government data sets, it is apparent that they provide substantial current and potential net benefits to the economy and society.
>
> In Australia, the estimated economic value of open government data sets range from a lower boundary of $500 million to an upper boundary of $25 billion per year. Globally, the potential value of open data (both public and private) could be up to $4 trillion per year. Significant benefits associated with open government data include improved government services, more efficient operations and business practices, better information exchange, and more engaged citizens, as shown by the sample projects and initiatives discussed in this report.
>
> ...
>
> Some of the potential high-value data sets held by governments that have been identified to date are spatial data, health data, transport data, mining data, environmental data, demographic and social data, and real-time as well as past emergency (e.g., bushfire) data (Department of Communications and the Arts, 2016).

Although the Australian reforms follow similar developments internationally, the Australian law continues to provide copyright protection over government material, unlike the legal situation in some jurisdictions such as under the US federal law. In Australia, there has at the time of writing been no change in the law dealing with copyright in government materials. Changes have occurred through the practice of copyright. As has been pointed out by other authors, the fact that government-produced information is in the public domain does not ensure access to government information.[16] Access to, and reuse of, government information requires policy, regulatory or other pro-active steps by government to ensure it takes place.[17]

The reforms can be, and are being, adopted in the presence or absence of copyright over government information and data.

4. The Future of the Australian Reforms

The efficacy of the reforms in Australia will be tested over time. There is an established momentum both in Australia and in many other countries and at an international level to open access to public sector information and data and this is likely to continue.

[16] Some agencies may have additional statutory authority to impose conditions for use. Reasons include ensuring that copyrighted information contained in the government product is recognized, adhering to agreements with other parties, and maintaining contact with users to ensure maintenance and updating of critical information. Refer CENDI "Frequently Asked Questions about Copyright: Issues Affecting the US Government" (October 2008) [3.1.5] http://www.cendi.gov/publications/04-8copyright.html#315.

[17] Gellman points out that the US federal government agency non-copyright information controls have included licence agreements and royalties, the limiting of access to selected recipients, denying or delaying access, agreeing with private companies to restrict access to data, hiding the data and restricting its use through contracts. Agency justifications for doing so have included data integrity and revenue raising. Robert Gellman, "Twin Evils: Government Copyright and Government Copyright-Like Controls over Government Information" (1995) 45 *Syracuse Law Review* 999.

In December 2015, the administering Department of Communications and the Arts released an exposure draft of the Copyright Amendment (Disability and other Measures) Bill 2016 for public comment. One proposal in the draft Bill is to have a fixed term for government material of 50 years from the date of making of the material regardless of it is published or not.[18] The proposal seeks to provide certainty over the term of copyright protection in dealing with both published and unpublished public sector information and data, and to that extent it is appropriate and helpful for those seeking access to that information, electronically or otherwise. While this is a modest reform, it is important for the future of Australian reform that copyright law and policy is consistent with, and not a barrier to, access to public sector information and data if the goals of accountability of government, better communication and interaction with the community and net benefits to the economy and society are to be promoted.

There have clearly been some proactive steps to facilitate access to, and reuse of, government information and data. A Data Portal is a useful aid to access and an important step. Access to, and reuse of, that information and data under the most liberal Creative Commons BY license is another.[19] Facilitation of access to, and reuse of government information and data would be further enhanced by a faithful commitment to the other federal public data policy requirements previously referred to, and in particular to make high-value data available for use by the public, industry and academia, in a manner that is enduring and frequently updated using high-quality standards. Another step is to add a quality framework to make it easier for users to identify high-quality reliable data.[20] A further step is to provide protection against legal action in making available information and

[18]A similar reform was proposed for non-government material.

[19]Material presented on data.gov.au is provided under a Creative Commons Attribution 3.0 Australia license.

[20]Australian Government. Department of Prime Minister and Cabinet, Data.gov.au www.data.gov.au/about. This Portal advises that "we are in the process of adding a quality framework to make it easier for you to identify high quality data you can rely on".

data to facilitate access to government data and information which may have originated from non-government sources. The government *Guidelines on Licensing Public Sector Information for Australian Government Entities*[21] contain examples of material unlikely to constitute PSI for the purpose of the Government's policy which include:

- Confidential material;
- Material that has national security and strategic interest implications;
- Material that contains personal information;
- Material that is commercially sensitive; and
- Material that is culturally sensitive.

The release of data containing this material is unlikely. Nevertheless, a regulatory proactive step by the government is to ensure that data and other information which may be released in good faith to the public that contains this excluded material should not be the subject of legal action. Similar provisions which have this effect presently exist in the *Freedom of Information Act* 1982 (Cth)[22] which protect the government, government officials and the author of a document from defamation, breach of confidence and infringement of copyright actions from the publishing or giving access to a document in good faith in the belief that publication or access is required or permitted. This includes the publication of accessed information to enable downloading from a website.[23] This legislative protection should desirably extend to the user of the data and information released by the government under these circumstances. Protection in this way would be a sensible and significant aid to public access to, and reuse of, government data and information.

---

[21]The *Guidelines on Licensing Public Sector Information for Australian Government Entities* slightly revised an earlier policy document, a draft version of which was released in 28 January 2011 and final version dated 28 February 2012: refer https://www.finance.gov.au/.../Draft-Guidelines-on-Licensing-Public-Sector-Informati....

[22]*Freedom of Information Act 1982* (Cth), ss 90–92.

[23]*Freedom of Information Act 1982* (Cth) s 11C.

Access to, and reuse of, public registry data and information raise similar issues. The *Freedom of Information Act 1982* (Cth) does not apply to documents which are open to public access.[24] These documents that become public on registration, such as in land titles registries, may contain third-party copyright material. The conditions under which documents are registered could expressly provide an open license for use by government and non-government users through, for example, a Creative Commons Attribution license (CC-BY),[25] or public access to, and reuse of, this information and data could be facilitated by legislative protection.[26]

More broadly federal and state Australian governments should provide proactive support, education and training to officials of government entities tasked with releasing open data, consistent with the introduction of Freedom of Information legislation in the 1980s across all Australian jurisdictions — and continue a dialogue with industry and the community in the development of new open data initiatives.[27]

References

Attorney General for Australia (2013). *Australia joins Open Government Partnership*. Media Release. Retrieved from http://www.attorneygeneral.gov.au/Mediareleases/Pages/2013/Second%20quarter/22May2013-AustraliajoinsOpenGovernmentPartnership.aspx.

Attorney General's Department (2010). *Statement of Intellectual Property Principles for Australian Government Agencies*. Retrieved from http://www.ag.gov.au/RightsAndProtections/IntellectualProperty/Documents/StatementofIPprinciples

[24] *Freedom of Information Act 1982* (Cth) s 12.

[25] Brian Fitzgerald, Anne Fitzgerald *et al.*, *Internet and E-Commerce Law, Business and Policy* (Lawbook Co, 2011), 431.

[26] Refer John S Gilchrist, *The Government and Copyright* (Sydney University Press, 2015) 217–218 for a discussion of these issues.

[27] Refer to the recommendations of Daniel Castro and Travis Korte, *Open Data in the G8: A Review of Progress on the Open Data Charter* (March 2015) (Centre for Data Innovation, Washington) 32 http://www2.datainnovation.org/2015-open-data-g8.pdf.

forAusGovagencies.pdf or http://www.ag.gov.au/RightsAndProtections/IntellectualProperty/Documents/IntellectualPropertyManual.pdf

Attorney General's Department (2012). *Guidelines for Licensing Public Sector Information for Australian Government Agencies.* Retrieved from http://www.ag.gov.au/RightsAndProtections/IntellectualProperty/Documents/GuidelinesforlicensingPSIforAusGovagencies.pdf.

Australian Law Reform Commission (2013). *Copyright and the Digital Economy: Discussion Paper* (DP 79), June 2013. Retrieved from http://www.alrc.gov.au/publications/copyright-and-digital-economy-dp-79.

Department of Communications and the Arts (2016). *Open Government Data — Why it Matters.* Bureau of Communications Research, February 2016, p. 33. Retrieved from https://www.google.com.au/?gws_rd=ssl#q=Open+Government+Data+-+Why+it+Matters'.

Department of Finance and Deregulation (2009a). *Engage: Getting on with Government 2.0.* pp. xvii. Retrieved from http://www.finance.gov.au/publications/gov20taskforcereport/.

Department of Finance and Deregulation (2009b). *National Government Information Sharing Strategy.* pp. 24–34. Retrieved from http://www.finance.gov.au/publications/national-government-information-sharing-strategy/docs/ngiss.pdf.

Department of Innovation, Industry and Regional Development (2010). *Whole of Victorian Government Response to the Final Report of the Economic Development and Infrastructure Committee's Inquiry into Improving Access to Victorian Public Sector Information and Data.* Parliament of Victoria, 2 February 2010. Retrieved from http://www.parliament.vic.gov.au/edic/inquiries/article/1019.

Department of Innovation, Industry, Science and Research (2009). *Powering Ideas: An Innovation Agenda for the 21st Century.* Retrieved from http://pandora.nla.gov.au/pan/109202/20091028-1042/www.innovation.gov.au/innovationreview/Documents/PoweringIdeas_fullreport.pdf.

Economic Development and Infrastructure Committee (2009). *Inquiry into Improving Access to Victorian Public Sector Information and Data: Report.* Parliament of Victoria, June 2009, Parliamentary Paper No. 198, Session 2006–2009. Retrieved from http://www.parliament.vic.gov.au/edic/inquiries/article/1019.

Productivity Commission (2016). *Intellectual Property Arrangements: Draft Report.* Canberra, Australia, April 2016. Retrieved from http://www.pc.gov.au/inquiries/completed/intellectual-property/draft.

The White House (2011). *The Open Government Partnership: National Action Plan for the United States of America.* 20 September 2011. Retrieved from http://www.whitehouse.gov/sites/default/files/us_national_action_plan_final_2.pdf.

Chapter 6

From Open Data to Open Governance in Canada: Dissecting a Work in Progress

Jeffrey Roy

School of Public Administration, Dalhousie University, Halifax, Nova Scotia, Canada

Many governments are striving to develop open data strategies, taking previously internal and often proprietary sources of information and rendering them public through online spaces. However, open data does not fit easily within the rubric of democratic governance and traditional public administration. The purpose of this chapter is to probe such tensions within the context of Canada, a Parliamentary democratic regime of the Westminster tradition where the inertia of the machinery of government often translates into a penchant for informational control rather than openness and sharing. More specifically, we examine the federal government's ongoing Open Government Action Plan and its three main dimensions: data, information and dialogue. Within each dimension, there are tensions between opportunities and pressures for openness and sharing on the one hand, and the inertia of traditional government and proprietary notion of information ownership and control on the other hand. Within a broader democratic context as well, notions of individual privacy coexist uneasily with the emerging culture of openness and sharing, a culture greatly facilitated by the advent of mobile computing and devices. This chapter concludes with a call for greater political innovation and dialogue in order to facilitate a more meaningful path of institutional adaption predicated upon enlightened openness and data sharing aligned with a culture of responsible and genuine public involvement in the creation of public value.

1. Introduction

Many governments are striving to develop so-called open data strategies, taking previously internal and often proprietary sources of information and rendering them public through online spaces. Such efforts are indicative of the emergence of 'meta-data' as a wider socio-economic and societal transformational driven by the advent of the Internet on the one hand, and massive and newly accessible data flows on the other hand (Harrison *et al.*, 2012; Helbig *et al.*, 2012; Ubaldi, 2013). These continually expanding data flows may well constitute new forms of public and private value if harnessed and exploited, something governments are now seeking to do via their open data efforts both directly (by opening up the information to wider access) and indirectly (by allowing for and stimulating innovative usage and application).

The movement partly draws sustenance from an evolving and increasingly open and data-driven society shaped by a confluence of both technological and governance patterns emphasizing mobility, collaboration and participation: it is underpinned by the backend dynamics of cloud computing for information storage and processing, and frontend Web 2.0 platforms (and social media in particular) and tools that facilitate content generation and sharing, thereby feeding the growing meta-data vortex. Open data is thus closely intertwined with the emergence of "Gov 2.0". However, open data does not fit easily within the rubric of democratic governance and traditional public administration (Roy, 2013, 2014a; Lips, 2016). Its emergence invariably encounters tensions between traditional public sector culture and structures more predicated on control-minded principles, such as secrecy and hierarchy (Aucoin, Jarvis & Turnbull, 2011; Grimmelikhuijsen, 2012; Roy, 2013).

The purpose of this chapter is to examine such tensions within the context of Canada, a Parliamentary democratic regime of the Westminster tradition where the inertia of the machinery of government often translates into a penchant for informational control rather than openness and sharing. At the same time, Canada is a federation — with open data and open government initiatives emerging first at the local level before being adopted federally. Accordingly, this chapter will draw selectively from all three levels of government (federal, provincial and local), with the predominant focus nonetheless at the federal level due to the resource intensity of this government on the one hand, and some important political changes that have been occurring through 2016 in light of recent federal elections toward the end of the previous year.

This chapter is thus organized as follows. Following this brief introduction, Section 2 provides some additional context for open data and open government — as well as the proprietary underpinnings of traditional government and the resulting tensions between two evermore competing worldviews of governance. Section 3 then provides a critique of open government in Canada — notably the federal government's Open Government Action Plan. Section 4 then extends this critique to a wider consideration of democracy in an increasingly data-driven world, and the tensions between proprietary notion of privacy and the challenges for a democratic culture that must balance privacy and individualization on the one hand, and widening calls for openness, participation and collective engagement on the other hand. A brief conclusion then draws together the key lessons from this undertaking for both Canada's public sector and governments more widely.

2. The Rise of Open Data and Proprietary Resistance

New participatory mechanisms, systemic openness and virtualization are underpinning an emerging governance ethos that, for the public sector, is often termed as the emergence of Gov 2.0. At the heart of Gov 2.0 are drivers of collective intelligence and more collaborative forms of governance that are typically associated with a widening online universe and less hierarchical and control-minded forms of governance (Shirky, 2008; Wyld, 2010; Maier-Rabler & Huber, 2011). From both external vantage points on new societal formations (such as Wikipedia and a myriad of social media-driven movements) as well as internal to the public sector (what Lips characterizes as "public administration 2.0"), governments are increasingly challenged to move beyond a typology of hierarchies and markets and embrace usage of networks typically more open and collaborative in formation and execution (Stoker, 2005; World Economic Forum, 2011; Kostakis, 2011; Gil-Garcia, 2012; Lips, 2012; Roy, 2013). Aligned with the spirit of such principles, the spreading of web 2.0 experimentation within government is specifically meant to foster collaboration and democratize the creation and exchange of ideas:

> The role of citizens in an open government environment — enriched by open government data — can be one of democratic innovators. In an ongoing open innovation process, citizens can draw on open data, and propose both policy-areas to tackle and technical approaches to take (Rabler & Hubler, 2011, p.186).

The potential recasting of governance in terms of expectations and roles is profound. Rather than gathering information and ideas via highly regimented and contained mechanisms (shaped by a proprietary mindset), this alternative presentation of openness and ideas begins from the premise that the ownership of information and ideas is fundamentally diffused and shared. At the same time, however, such an ethos of openness invariably faces strong pushback from both the traditions of proprietary protection and its organizational cousin that is particularly prevalent in the public sector — namely hierarchical and informational control (Grimmelikhuijsen, 2012; Roy, 2013). For example, one early study of the usage and acceptance of new social media within the public sector found such tensions deeply engrained within Canadian government where information is viewed predominantly as a proprietary asset. The authors conclude that the most significant impediment to Gov 2.0-inspired reform is the "clay layer" embedded by a hierarchical public service culture (Fyfe & Crookall, 2010, p. 3).

By contrast, open government is based upon the "notion that public sector information is a resource, the release of which will maximize its social and economic value to citizens" (*ibid.*). In the Netherlands, for example, an impetus for non-proprietary public data came from the Dutch courts in April of 2009 when a City of Amsterdam's appeal to impose restrictions and fees over several its data holdings was rejected (Ubaldi, 2013). Such clashes between proprietary and openness, and control and empowerment shape the pursuit and effectiveness of open data and its wider ramifications (Bermonte, 2011; Roy, 2013). Outside of government too, similar tensions between proprietary and open systems are prevalent across many segments of industry and society (Wyld, 2010; Public Administration Committee, 2011; World Economic Forum, 2011).

Yet a widening ethos of openness draws sustenance from: (i) the Internet as a platform for democratization in the broadest sense; (ii) the search engine and a widening array of self-expressive and interactive web 2.0 tools and platforms; and (iii) most recently the advent of mobility. As Young puts it, the cloud as a symbolic basis of a wider virtual universe driven by a myriad of smaller and more powerful and mobile computing devices, a penchant to share more and more personal information online — especially via social media, and a new form of enhanced and shared networked intelligence (Young, 2012). At the same time, however, accompanying optimistic portrayals of the potential benefits of such intelligence come offsetting concerns pertaining to individual privacy, while open data has similarly sparked fresh concerns about the digital divide and accentuating new forms of "data divides" (Halonen, 2012).

Indeed, as important to government efforts to release data is society's interest and ability in accessing and making use of it. Open data's origins are interwoven with a growing community of activists and apps developers working initially within the confines of privately-developed operating system platforms such as Apple and

Android (the latter built from open sourced coding and thus more portable across a range of companies and devices). The participative flavor of such movements can and has also extended beyond commercial pursuits, as exemplified in February 2013 by the inaugural open data day (the featured a global hackathon of events in cities around the world).

In this vein, Washington D.C.'s local government would pioneer in 2009 what is believed to be the world's first public sector apps competition, deemed "apps for democracy". In Canada, the City of Edmonton would be the first Canadian jurisdiction to adapt such a model, and the first to devise a formal open data and open government strategy (Roy, 2014). Today, such initiatives are commonplace across the country with the federal government devising its own Open Government Action Plan as a blueprint for its own likeminded efforts.

3. A Critique of Open Data and Open Government in Canada

Within the context described above, we now seek to both dissect and examine open data and open government in Canada, doing so through the lens of three inter-related dimensions: data, information and dialogue. These dimensions are adopted from the three main components of the Government of Canada's Open Government Action Plan (albeit a plan formulated under the previous political, Conservative regime defeated in federal elections in late 2015). Additionally, it should be recalled that Canada is a federation and so the distinction between federal and national is important, the latter encompassing both provinces and the federal government as well as local governments that all have their own and often separate open data and open government plans.

3.1. *Data*

As of June 2016, the Government of Canada's open data portal featured more than 120,000 data sets gathered from various federal departments and agencies — and readily available under an open usage license now adopted by many provinces and local jurisdictions as well. Such a license allows anyone to "copy, modify, publish, translate, adapt, distribute or otherwise use the information in any medium, more or format for any lawful purpose". The portal and its open data mission are based upon three underlying principles: (i) that government data should be easy to reuse and in open formats; (ii) that government data should be easily discoverable including via federated search; and (iii) that government should engage with citizens, especially to prioritize data for release.

Despite significant and widening open data experimentation in terms of the volume of data sets released by governments at all levels in Canada, the main challenge facing all jurisdictions is that of leveraging such data openness into opportunities for public value creation on the one hand, and demonstrating and measuring such public value creation on the other hand. Aside from showcasing in general terms online the sorts of data sets most widely searched — and a sample of the sorts of apps that have been created by government-sponsored apps contests across the country — there is little in the way of specific performance data illuminating to what extent such data resources are being used, by whom, and for that purpose. Accordingly, data availability alone — though a laudable first step and an important source of experimentation — falls well short of devising capacities for "ubiquitous engagement" sought by enthusiasts of systemic openness and more participative and collaborative forms of governance (Lee & Kwak, 2011; Roy, 2013, 2014; Lips, 2016).

Canada is hardly alone in this regard as Lips explains from the New Zealand vantage point:

> There has been hardly any evidence thus far of innovative collaborative partnerships between government agencies and the general public, let alone between government agencies and different stakeholder groups of the New Zealand population, including Māori, Pakeha4 and Pasifika. In fact, the mostly viewed datasets, such as the statistical profiles of New Zealand Councils, New Zealand import and export statistics, and the Christchurch earthquake land

> check colour zones, suggest that members of the general public perceive opened up government data not as an 4 Pakeha is the Māori name for New Zealand Europeans 17 opportunity for achieving citizen empowerment or collaborative democracy through government data (re-)use and innovation, but as a particular form of digitised government information provision to the public (Lips, 2016, p. 16).

It should nonetheless be noted that the Government of New Zealand is much more advanced in their development and releasing of performance reporting with respect to open data than is the case for the Government of Canada (which limits itself to some basic bullet points on the open data portal sketching out the number of data sets and the most widely accessed levels). Indeed, while it is admittedly hard to understand how Canada fares fourth in the 2015 Open Data Barometer (ranking high on readiness and implementation), the country's lowest score from this ranking scheme does apply to impact (where political and economic impacts fare especially poor relative to social impact). Similar findings in Canada have been observed at the local level where municipalities fare generally much more poorly in gauging and demonstrating impacts than with respect to readiness and implementation (Open Cities Index, 2016).

While a plethora of studies and articles has been devoted in recent years to the technical and legal architectures of open data (issues such as data formatting and cleanliness and licensing restrictions), the fundamental constraint on such programs is arguably their novelty and the cultural and structural impediments to openness reflected in the wider governance and democratic environments. This observation is reflected in the Open Data Barometer itself and the sharp regional differences between developed and developing countries (and the close correlation between open data and wider governance transparency and digital government progress as measured by other global studies such as those of Transparency International and the United Nations). It is also reflected within developed countries as well: in one major survey in the US, for example, PEW Research found that Americans were poorly informed about open data opportunities and generally skeptical about their value in light of general distrust toward the public sector both operationally and politically (PEW Research, 2015).

In sum, while Canada can claim significant progress on data availability (not only the federal open data portal but similar like-minded initiatives provincially and locally), there are significant and unanswered questions remaining as to what if any impacts such data availability is generating — and there are wider questions still as to whether open data is aligned within a more ambitious rubric of open government. It is these latter questions that are our primary interest in this chapter, thus turning to the second and third dimensions of the federal government's Open Government Action Plan (information and dialogue).

3.2. *Information*

Governments of any political stripe are generally more comfortable with the notion of sharing raw data than sharing information on their policies and decision-making practices and processes. Indeed, within the Westminster political model, secrecy is at the core of how Governments function and central agencies are an extension of this mindset. It should not go unnoticed, then, that within the Government of Canada, the Open Government unit responsible for open data and likeminded initiatives resides within the Treasury Board Secretariat (TBS), one such traditional central agency known more for its monitoring and control-minded functionality across government (especially in terms of spending and operations) than for its enthusiasm for openness. In fairness, this organizational base does allow the federal Open Government Action Plan to recognize, at least rhetorically, the importance of information along with data (the third dimension being dialogue, as noted). Accordingly, the previous federal government sought to showcase reforms in the realm of more information openness such as a new online portal for information requests and better website capacities enabling more integrated searches for government information via both desktop and mobile devices.

Nonetheless, a wide consensus exists that the previous government's commitment to information openness was far from genuine (Martin, 2010; Aucoin, Jarvis & Turnbull, 2011; Roy, 2013). Such concerns have persisted under the new Liberal regime — which had promised to fundamentally reform the access to information regime and to significantly bolster transparency. Six months into their mandate, the Information Commissioner has already sought to raise alarm bells about backsliding on such a promise, with the Parliamentary Committee's most significant findings and proposals deferred until 2018 at the earliest. It obviously remains premature at the time of writing to judge the new government which has nonetheless also been credited with encouraging a cultural shift within the ranks of the federal public service (especially in allowing scientists to speak more freely to media and outside stakeholders, a practice that the previous government had sought to contain and carefully orchestrate through its political communications apparatus).

Indeed, the significant challenge confronting the new government is in itself a familiar one within the context of traditional public administration and representational democracy — namely a culture of information control that itself is rooted in the very foundations of decision-making authority, notably a political cabinet and central agencies. This latter group of bodies is especially relevant to the emergence of open data and open government, a point exemplified by the housing of the unit responsible for the Open Government Action Plan within the TBS. Treasury Board's primary mission, to exert expenditure review and monitoring on almost all federal department and agencies, exudes a control-minded culture directly at odds with the ethos of openness sought by advocates of reform. It does bear noting that the reasoning for this seemingly counterintuitive positioning is that the open government unit resides within the Office of the Chief Information Officer (CIO), itself a body within the wider TBS structure.

Furthermore, the rationale for this CIO placement is to ensure government-wide coverage in the pursuit of infrastructure and policies to facilitate interoperability and information sharing within the federal apparatus. Yet this inward focus on infrastructure planning and information management frameworks has been historically shaped by not only the control-laden values of traditional public administration but also an accompanying proprietary ethos of industry perspectives on digital government that in many ways reinforce this culture of secrecy and control (Public Administration Committee, 2011, p. 12):

> We recognise that there will be resistance to this approach. Governments have traditionally limited their ability to publish this information by signing commercial confidentiality agreements with companies. In future such agreements must be severely restricted to enable the Government to publish detailed contractual information about how much they are paying for different services and products within a contract. This should disadvantage nobody if all suppliers are treated the same.

At the federal level in Canada, beyond some experimentation with open data there is little evidence on such an ethos of openness being pursued more deeply in terms of wider approaches toward information governance and a greater emphasis on openness. As one recent study of digital government in Canada framed it, such an ethos of openness is indeed contradictory toward how the public service has itself functioned for some time and what is called for is a reframing away from hoarding and specialization to information sharing as a core professional value:

> Too often, data in the public service is viewed in terms of scarcity, as a resource that increases the worthiness of individuals and ministries by virtue of being owned and controlled. When data is viewed as proprietary, rather than as a corporate or public good every step towards transparency and integration will be long, fraught with obstacles and expensive. Government needs to re-examine performance planning, promotion, budget-granting, and hiring practices to recognize and rewards 'sharers' rather than 'hoarders' (Johal & Galley, 2014, p. 41).

A government therefore struggling to share information both within itself and more widely with the external citizenry — embracing transparency in a reactive and trepid manner — is thus one that is unlikely to be well-suited to the third pillar of its Open Government Action Plan: dialogue.

3.3. *Dialogue*

As is the case in much of the world, governments at all levels in Canada routinely make pronouncements about linking openness to wider opportunities for consultation and engagement. While this emphasis on public participation predates the digital age, it has benefited from a huge impetus since the advent of the Internet and more so, models of Gov 2.0 emphasizing "ubiquitous engagement" as the key to public value creation (Lee & Kwak, 2011).

The Lee and Kwak formulation from the American vantage point is notable in its direct linkage to open data and open government as a platform for wider engagement, a viewpoint stemming from the inaugural Presidential Directive in 2009 by the Obama Administration on openness which embraced three inter-related principles: transparency, participation and collaboration. Although the Obama Administration has clearly not been without its own challenges within the realm of information — and especially strong tensions between information sharing and classified information sources — evidence also exists to support the notion that the US federal government has sought to encourage new forms of online participation and engagement (Lee & Kwak, 2011; Mergel, 2012). There is also a strong tie between the 2009 Obama Directive and the international Open Government Partnership (OGP) that has sought to globalize these principles and enshrine them into other national governments and their efforts (with mixed success as denoted by the Open Data Barometer and other likeminded sources).

In Canada, the Conservative Government's inclusion of dialogue as a core dimension of its Open Government Action Plan reflects this globalizing logic and the close association between openness and online empowerment and engagement. Under the Action Plan, the following commitments were made:

- Improve the existing Consulting with Canadians website to facilitate easier access to information on federal consultation activities for citizens.
- Develop and launch a new government-wide consultation portal to promote opportunities for public participation, host online consultations, and share findings from completed consultations.
- Expand the use of social media across government to enable departments and programs to connect to Canadians in innovative ways and enhance engagement in support of citizen-centric services.
- Develop a set of principles and standards for public consultations in discussion with citizens and civil society (e.g., advance notice and promotion of consultations, best practices for in-person and online engagement, effective use of social media, reporting on results), including setting out minimum benchmarks for consultations.
- Conduct targeted consultations on open government themes with key groups in Canada (e.g., youth, Aboriginal populations) (Government of Canada, 2014–2016).

Despite the laudable rhetoric, much more than is the case for information and data, the government's own reporting on its own efforts with this dimension of dialogue proved to be completely devoid of any tangible realization of progress — or even much in the way of experimentation for the sorts of directions described above. Such inaction is rooted in the wider tension between the Westminster model of Parliamentary Government (including federal and provincial governments in Canada — though excluding municipalities whose structures are nonetheless determined by provincial legislation) and its penchant for informational control, versus the ethos of open government and open governance emphasizing information sharing and engagement (Roy, 2013).

Aside from this rather bleak characterization of stunted dialogue, there are some notable exceptions both in the past and present. For instance, the previous government sought to create an apps competition from its open data platform (following the lead of various local and provincial governments as well as other jurisdictions elsewhere), arguably a form of dialogue in encouraging third-party apps developers to engage in this form of crowd-sourced innovation (whereas average citizens without the skills to design apps were encouraged to submit and discuss concepts for apps with a potential for creating public value). Moreover, the Open Government team housed within the Treasury Board also sought to engage key stakeholders and the public at large in a consultative exercise to infuse the Open Government Action Plan with specifics. The most recent phase of this ongoing effort took place in the spring of 2016 with an open invitation for feedback on the action plan and where it could potentially be headed next.

With respect to social media — a key aspect of online dialogue for any government — there is empirical evidence to support the notion that the previous government viewed social media as predominantly a communications platform for informational purposes both administrative and political — despite the rhetoric above with respect to commitments for leveraging social media as an interactive forum for greater citizen involvement. For instance, a detailed review of social media practices — and the culture within the federal government underpinning such practices — revealed that public servants themselves felt highly contained in their usage of social media tools — and with respect to their freedom to express themselves openly on matters of their own expertise, a particular point of contention in science-intensive fields such as agriculture, health and the environment (Harrop, 2016).

By contrast, the new Liberal Government has publicly committed to a recasting of this approach and to a greater usage of social media and new technologies to expand citizen involvement in governance matters both in terms of policy-making and service delivery: such themes were central elements of the Ministerial mandate letters released for the first time in Canadian federal government history in the fall of 2015. Here, however, an important distinction must be made between the administrative apparatus of the state (i.e., the public service) and the political wing, despite both co-existing within the executive branch. With respect to the public service, the government was relatively quick to set a new cultural tone by encouraging public servants — notably scientists — to speak and express their viewpoints openly via traditional and new media, potentially incentivizing the sort of directions envisioned by the previous version of the Open Government Action but never realized in practice.

Politically, the first key test of the new government's appetite for novel and expanded forms of public dialogue is the democratic reform file — notable in part for the appointment of a young minister with an explicit mandate to engage in such a broad discursive effort potentially recasting the functioning of political institutions in important ways:

> As Minister, you will be held accountable for our commitment to bring a different style of leadership to government. This will include: close collaboration with your colleagues; meaningful engagement with Opposition Members of Parliament, Parliamentary Committees and the public service; constructive dialogue with Canadians, civil society, and stakeholders, including business, organized labour, the broader public sector, and the not-for-profit and charitable sectors; and identifying ways to find solutions and avoid escalating conflicts unnecessarily (Office of the Prime Minister, 2015).

Yet the broad contours of this mandate and its potential were quickly shaped by a near all-encompassing focus on electoral reform — and the political antics of Parties in Parliament to form a Committee that would review various electoral models and provide guidance to the government. The ensuing dynamics quickly became politically contentious on matters such as the composition of the Committee, varying opinions as to whether a referendum should be used to seek a final mandate for electoral change, and a very tight timeline for public consultation prior to the Committee formulating its positions (leading many in opposition to suspect that the government is merely seeking to craft an endorsement of its own preferred option). Despite the minister's suggestion that social media and other online capacities would be central to this consultative effort, the tight timeline dictating limited opportunities over the summer months for public dialogue and input have been met with suspicion by many groups, and largely indifference by the public at large (at least at the outset of this summer process).

The result of such dynamics and constraints is that the entire exercise of democratic reform has quickly come to resemble a case study in closed and traditional government — rather than a model of newer and more outward and open governance. With respect to the future of open government — and what it might mean for citizen dialogue and active involvement in public sector governance — a deeper consideration of the relationship between data, privacy and openness, and democratic culture is called for.

4. Data and Democracy

The pursuit of openness and more participative forms of public value creation merits a more direct consideration of the impacts of widening pressures for openness and sharing on democratic conduct and culture, particularly in a more data-centric environment where tensions between openness and secrecy of organizations (both public and private) are themselves mirrored by an evolving individual calculus of personal privacy and digital security. In other words, an online, virtual and increasingly mobile ethos entails how individuals behave across both private and public pursuits and whether and how such behavior impacts the formalities of political activity and democratic accountability.

As the public seeks benefits from online activity in numerous respects ranging from consumer purchases to social media communities themselves premised upon business models driven by commercial advertisement, more and more personal information is readily made available and shared, a dynamic propelled by the spreading of a mobile internet (Young, 2012; Roy, 2013; Wood, 2016). Indeed, stemming from this engrained logic of commercialization and customer service often embraced and deployed by governments themselves in shifting services online, the public is routinely assured that their privacy shall be respected. Partly due to such assurances and partly stemming from an over-riding ethos of convenience, the default decision for many individuals online is to click acceptance of a privacy policy that is neither read nor understood (*ibid.*).

There are three salient and potentially contradictory points with respect to the interface between online privacy and democracy. First, driven by an online ethos of openness and sharing on the one hand, and a commercialized logic of convenience and gratification on the other hand, individuals are encouraged either directly or indirectly to share more of their personal information online. Second, any understanding or fundamental accountability of how such information flows are gathered and used is arguably quite limited and opaque. Third, and most vitally, if citizens are not prepared to demand transparency and answerability from private sector companies (often housing the platforms of online commerce and information sharing), then they are unlikely to do so from their governments — either in terms of how governments gather and use their own personal information or whether and how governments gather such data from the private sector, as is increasingly the case (Roy, 2016).

By contrast, the necessity of more active forms of public awareness, vigilance and involvement (and by extension the escalating risks of passivity and ignorance) can only augment going forward, in what Wolf describes as the "data-driven life" emerging from the interplay of four fundamental alterations to the relationship between people, data and technology that is driving the emergence of the cloud (Young, 2012, p. 11). As Young explains, current approaches to managing data are poorly suited to this new environment: "the systems we have in place around contracts and consent and rights are really designed for an analog era, an older informational ecosystem" (*ibid.*, p. 183).

Part of her proposed basis for new solutions — and a preservation of the "digital self" — is an emphasis on the value and importance of personal responsibility. She thus calls upon individuals to become "data activists" in moving beyond passive usage and acceptance of new informational offerings that bundle, share and integrate data in incomprehensible and seemingly opaque manners, and to seek greater openness, understanding and vigilance in what amounts to an individualized determination of appropriate privacy trade-offs on the one hand, and greater degrees of collective enlightenment on the other hand. Such a reorientation of privacy as much about responsibility as rights provides some basis for finding middle ground between the extremities of those seeking outright but increasingly unattainable privacy guarantees and those instead viewing privacy as unrealistic or simply out of date. It can also draw sustenance from the collectivized dynamics of open source communities that have sought some degree of balance between open collaboration and competition in fostering new innovations in both backend infrastructure and content creation.

Yet such calls are widely out of step with our individualistic and legalistic portrayal of privacy as an enshrined right, even one routinely circumvented by governments themselves in the name of security and — paradoxically — thwarting other government efforts to integrate and improve service offerings and the customer experience (Roy, 2013). Consequently, the significant danger of a large rights and reassurance orientation toward privacy matters is an accentuation of passiveness with respect to individual awareness and behavior, a conditioning that arguably extends to political matters more generally (*ibid.*).

A new form of digital literacy is thus called for. Closely tied to notions of digital inclusion but extending to even those already technologically adept, the central meaning and purpose of digital literacy is to build upon Young's call for data activism and instill an ethos of enlightened technological usage on the part of the individual, as well as an expectation of corporate, public and collective accountability in terms of organizational conduct is an over-more data-centric and networked era. A 2016 survey of Canadians' significant and widespread propensity to use public wifi networks with little or no regard to the absence of privacy safeguards further underscores the urgency of such issues: the survey found that more than two-thirds of

Canadians believe their information is "somewhat" or very safe when using public wifi, the highest level of the nine countries surveyed (Wood, 2016).

The rapid proliferation of mobile devices thus greatly enhances the urgency of this cultural rebalancing, as devices such as smartphones and tablets are increasingly the gateway of choice for individuals to access social media and cloud computing platforms. Notions of duty and responsible usage should also have a place in democracy, not a particularly weighty assertion by historical measure but one increasingly out of step in more recent times with the rise of the predominantly commercial dimensions to online infrastructure and the manner by which mobility is accentuating the pursuit of often individualized economic value at the expense of more collectivized formations of public value creation. Within democracies, then, digital literacy must be tied to civic literacy and the need to address what Nabatchi has termed as the democratic deficit ailing our collective institutions (Nabatchi, 2010). In other words, mobility's spreading must bring with it not only opportunities for enhanced service and participation but also an expectation of engagement that ties personal data activism and civic responsibility more broadly. Such a cultural rooting underpins any meaningful cocreation of public value between the citizenry and the public service within an open, responsive and deliberative democratic architecture.

Consequently, if open data and open government initiatives are to be aligned with more open forms of democratic governance, the contradictions of the newly emerging participatory ethos must be both recognized and addressed. Openness encourages and is tied to a widening digital culture of sharing — both by governments and by individuals, albeit with some uneven recognition and application of proprietary concepts and restrictions. Yet there is a widening disconnect between traditional government, where openness exists mainly on the margins rather than at the core of the decision-making apparatus, and the rhetoric of open government that espouses greater openness and outward engagement are core operating principles. The resulting participative ethos is thus highly limited to niche opportunities such as data-driven innovation (i.e., apps contests and the like) and service improvement processes driven by a customer-centric view of governance where the public sector is both implicitly and explicitly compared and measured to the commercial sector.

Accordingly, if openness is to afford a new participative culture of "ubiquitous engagement" (Lee & Kwak, 2011) where citizens are more genuinely involved in the cocreation of public value, a term which can have both administrative and political meanings as Stoker and others explain (Stoker, 2005), a culture of democratic openness must be paired and aligned with a culture of enlightened engagement which begins with a dialogue as to how to adapt democratic institutions for an era where sharing and networked connectedness rival and in some cases surpass the centrality of individualistic notions of privacy as an enshrined and assumed right (Roy, 2016).

In a select number of democratic jurisdictions also widely regarded as digital leaders in many respects of public sector reforms, we can see at least the makings of such a dialogue and some of its consequences. In Estonia, for instance, an online and mobile-friendly service delivery apparatus integrates and shares information across many public and private agencies while also providing citizens with the ability to track who is accessing their data and for what purpose while in Scandinavian countries, tax records of individual citizens are both pre-completed by state authorities (facilitated by information sharing amongst public and private agencies), and then fully released and accessible online, providing an unprecedented and controversial level of systemic transparency between citizens and government (that would be unthinkable in many Western countries and certainly in a North American context that continues to cling to the notion that individual privacy remains both appropriate and attainable as a guiding principle and pillar in a digital world) (The Economist, 2016).

Although distinctions across political cultures and democratic systems are multi-faceted, it is interesting to observe that while the US and the UK rank highest on a reputable global survey of open data capacities (countries paradoxically known for their individualistic and privacy-centric market and political cultures), they drop below the aforementioned Scandinavian countries on broader measures of openness and transparency such as the perception of public sector corruption as measured by Transparency International. This result may suggest that over time, countries such as the US and the UK (and by extension Canada and Australia) may have greater difficulty leveraging the fuller benefits of open data and open government, due to more traditional political structures that also translate into a more traditional and adversarial climate with respect to individual privacy — as has been demonstrated by the 2015 dispute between the US federal government (notably the FBI and the Department of Justice) and Apple computers over the unlocking of an iPhone of a terrorist suspect. Indeed, this hypothesis finds some support in a recent survey of Americans by PEW Research that found widespread distrust of government's information and data openness — and relatively little usage of such resources by the citizenry (PEW Research, 2015).

The stark contrast between this latter study and the US's position as number two on the aforementioned global survey of data openness (opendatabarometer.org) underscores the contradictory forces at play. It also aligns with Lips's findings in New Zealand regarding the limited impacts of open data in that country due to the cultural limitations both within government and across the public at large (Lips, 2016). Such findings further resonate in Canada where there has been little effort undertaken by the government to demonstrate the sustainable impacts of open data strategies, and where there has been the sorts of administrative and political resistance to wider openness discussed throughout this chapter. While open data resources are undoubtedly being leveraged by limited segments of society such as apps developers, researchers and specialized stakeholder groups, there is little evidence from Canada that they have sparked wider cultural and structural changes either within the public service or in terms of the interactions between the public service and the citizenry. One key reason for this limitation is that the federal government is viewed as primarily traditional in viewing information and dialogue through a proprietary and control-minded lens on the one hand, while also continuing to espouse the upholding of fundamental privacy rights (even as the government routinely overrides such rights in ways that are acknowledged generally, such as national security matters, but otherwise shielded from public purview) for individual citizens.

At a minimum, then, we can say that if open data and open government are to become enablers of a culture of greater democratic openness (seemingly a goal sought by more and more governments around the world, and one in keeping with the trends of a more informed and potentially engaged citizenry), a political conversation is required as to the appropriate balance between openness and sharing individually and collectively on the one hand, and the implications of such openness and sharing for both personal privacy rights and responsibilities — including, importantly, the responsibility to contribute to a more participative ethos of democratic governance. Such participation can be relatively passive in form — and focus predominantly on caring for one's own information and data and understanding its usage in increasingly networked processes, or it can be more active in form — by way of more direct engagement in policy and service innovation processes established to cocreate public value.

5. Conclusion

In Canada, as well as its likeminded political cousins and neighbors (notably the UK, US, Australia and New Zealand), open data undertakings by public sector authorities are an important recognition of the emerging ethos of Gov 2.0 predicated upon seeking collective benefit from more outward, open and networked forms of governance. Yet they have done relatively little to fundamentally alter the democratic institutions of these countries, greatly limiting the overall impacts of such efforts. While this assessment varies across countries, this chapter has sought to demonstrate the inertia of traditionalism within the Canadian context, while also situating the Canadian experience within a larger and more comparative lens.

In her own examination of the New Zealand context with a similar tone of gradualism and traditional inertia stemming from a rigorous review of open data efforts in that country, Lips concludes that the new "ideal state" of open governance is not likely to come about from changes rooted within the public sector but rather by pressures and dynamics emerging outside of it (Lips, 2016). She also underscores that in "attempts towards public sector reform technology only plays a very minor role: if a pathway is determined in this long, slow and evolving process of public sector change, it is by institutions, public managers and political leaders" (*ibid.*).

This latter point seems essential in that political leaders are the linchpin between the public service executive branch and the citizenry and at large, and the sorts of externalized pressures of which Lips speaks. Accordingly, a high degree of political literacy in digital matters generally and with respect to openness specifically is likely to be a central element in any meaningful renewal of the democratic institutional landscape going forward. In Canada, such political leadership has been observed in modest degrees at local levels, where open data initiatives were first pioneered, but they have yet to translate into a more robust and activist presence by elected officials on digital matters provincially and federally. In the UK, open data and open governance have arguably benefited even modestly from the workings of Parliament on such matters and a high level of political support for digital government in recent years, including a Digital Democracy Commission rooted within the Parliament itself (though such political engagement and support are now seemingly at risk in light of the newly emerging Brexit dynamic taking hold in that country).

In sum, data innovation and political innovation are intertwined in shaping a new and expanded ethos of openness and participation, or in limiting it within the confines of both traditional government and democracy on the one hand, and traditional notions of individual privacy on the other hand. Going forward, the co-creation of public value in an increasingly virtual and networked world necessitates new discursive mechanisms with the citizenry forged upon a rethinking of data ownership and civic participation in this new era.

References

Aucoin, P., Jarvis, M. & Turnbull, L. (2011). *Democratizing the Constitution: Reforming Responsible Government*. Toronto: Emond Montgomery Publications.

Bermonte, A. (2011). Senior leaders' use of web 2.0 and social media in the Ontario Public Service. Theses and dissertations. Paper 680. Ryerson University, Toronto. Retrieved from http://digitalcommons.ryerson.ca/dissertations/680.

Gil-Garcia, J. R. (2012). *Enacting Electronic Government Success: An Integrative Study of Government-Wide Websites, Organizational Capabilities, and Institutions*. New York, NY: Springer.

Government of Canada (2014–2016). Canada's Action Plan on Open Government. Retrieved from http://open.canada.ca/en/content/canadas-action-plan-open-government-2014-16.

Grimmelikhuijsen, S. (2012). A good man but a bad wizard. About the limits and future of transparency of democratic governments. *Information Polity*, *17*, 293–302.

Halonen, A. (2012). Being Open about Data Analysis of the UK open data policies and applicability of open data. Finnish Institute, London. Retrieved from http://www.finnish-institute.org.uk/images/stories/pdf2012/being%20open%20about%20data.pdf.

Harrison, T. M., Guerrero, S., Burke, B. G., Cook, M., Cresswell, A., Helbig, N., Hrdinova, J. & Pardo, T. (2012). Open government and e-government: Democratic challenges from a public value perspective. *Information Polity*, *17*, 83–97.

Harrop, L. (2016). Under the Microscope: Exploring Government's Use of Social Media in Communicating Science. School of Public Administration, Dalhousie University, Halifax, NS.

Helbig, N., Cresswell, A. M., Burke, B. G., Luna-Reyes, L. (2012). The Dynamics of Opening Government Data (A White Paper). Center for Technology in Government, Albany. Retrieved from www.ctg.albany.edu.

Johal, S. & Galley, A. (2014). Reprogramming Government for the Digital Era. Mowat Centre, University of Toronto, Toronto.

Kostakis, V. (2011). The advent of open source democracy and wikipolitics: Challenges, threats and opportunities for democratic discourse. *An Interdisciplinary Journal on Human in ICT Environment*, *7*(1), 9–29.

Lee, G. & Kwak, Y. (2011). An Open Government Implementation Model: Moving to Increased Public Engagement. IBM Center for The Business of Government.

Lips, M. (2012). E-government is dead: Long live public adminstration 2.0. *Information Polity*, *17*, 239–250.

Lips, M. (2016). Opening up government-held data in New Zealand: Implications for governance in the digital age. In *20th International Research Society on Public Management Annual Conference*, Hong Kong, 13–15 April.

Maier-Rabler, U. & Huber, S. (2011). "Open": The changing relation between citizens, public administration, and political authority. *eJournal of eDemocracy and Open Government*, *3*(2), 182–191.

Martin, L. (2010). *Harperland: The Politics of Control*. Penguin Group Canada.

Mergel, I. (2012). The social media innovation challenge in the public sector. *Information Polity*, *17*, 281–292.

Nabatchi, T. (2010). Addressing the citizenship and democratic deficits: The potential of deliberative democracy for public administration. *The American Review of Public Administration*, *40*(4), 376–399.

Office of the Prime Minister (2015). Minister of Democratic Institutions Mandate Letter. Retrieved from http://pm.gc.ca/eng/minister-democratic-institutions-mandate-letter.

Open Cities Index (2016). Open Cities Index Report 2015. Public Sector Digest. Retrieved from https://www.publicsectordigest.com/articles/view/1547.

PEW Research (2015). American's views on open government data. Retrieved from http://www.pewinternet.org/2015/04/21/open-government-data/.

Public Administration Committee (2011). *Government and IT — "A Recipe For Rip-Offs": Time for a new approach*. Twelfth Report of Session 2010–2012. Parliament of Great Britain, London.

Roy, J. (2013). *From Machinery to Mobility: Government and Democracy in a Participative Age*. New York, NY: Springer.

Roy, J. (2014). Open data and open governance in Canada: A critical examination of new opportunities and old tensions. *Future Internet Journal*, *6*(3), 414–432.

Roy, J. (2016). Mobility and service innovation: A critical examination of opportunities and challenges for governments in Canada. *International Journal of Public Administration in the Coming Age*, *3*(4), 1–14.

Shirky, C. (2008). *Here Comes Everybody: The Power of Organizing Without Organizations*. New York, NY: Penguin Group.

Stoker, G. (2005). Public Value Management — A new narrative for networked governance? *American Review of Public Administration*, *36*(1), 41–57.

The Economist (2016). Two rights, wrong policy. Retrieved from http://www.economist.com/news/leaders/21696939-push-publish-peoples-tax-returns-pits-transparency-against-privacy-which-should-win-two.

Ubaldi, B. (2013). Open Government Data: Towards empirical analysis of open government data initiatives. OECD Working Papers on Public Governance, No. 22, OECD Publishing. Retrieved from http://dx.doi.org/10.1787/5k46bj4f03s7-en.

Wood, E. (2016). Norton discovers Canadians rust public Wi-Fi too much, releases app to help. Retrieved from http://www.itbusiness.ca/news/norton-discovers-canadians-trust-public-wi-fi-too-much-releases-app-to-help/74507.

World Economic Forum (2011). The future of government: Lessons learned from around the world. Global Agenda Council on the Future of Government, World Economic Forum.
Wyld, D. C. (2010). Moving to the cloud: An introduction to cloud computing in government. IBM Center for the Business of Government.
Young, N. (2012). *The Virtual Self*. Toronto: McClelland & Stewart.

Chapter 7

Toward Open Innovation and Data-Driven Health Policy Making

Marika Iivari, Minna Pikkarainen*,†, Julius Francis Gomes*, Jukka Ranta† and Peter Ylén†*

**University of Oulu, Finland*

†VTT Technical Research Centre of Finland, Finland

Addressing digital open innovation from the policy perspective, we explore how the healthcare policy makers, concerned with producing more effective and preventive policies, would benefit from more open processes of knowledge flows. What is the role of data in health policy making? What could be done to advance open innovation in the healthcare sector based on heterogeneous sources of data? Digital technologies have enabled new ways to generate, collect, analyze and share health-related data, which has greatly contributed to opening the healthcare sector. The underlying thinking is that more readily available information, together with policies aimed at prevention rather than treatment, could radically reduce the costs in healthcare, while simultaneously improving the quality of care. However, healthcare as an innovation system is very different from other sectors, setting specific challenges on knowledge sharing and processes. Based on empirically grounded research in Northern Finland with a focus on preventive data-driven healthcare, this chapter studies how healthcare policy making could utilize more distributed data, information and knowledge flows across organizational boundaries. The results of the study suggest that technological and analytical solutions brought by digital technologies have the ability to support faster and better use of data in decision-making, and speed up the distribution of knowledge flows across different public decision-making bodies, and break the silos in public policy making for producing better and more efficient health and well-being services for the citizens. However, the use of different data sources is effective in preventive healthcare only if knowledge is systematically integrated and carried across organizational boundaries. This highlights the central role of policy makers as the leaders in advancing open innovation in the public sector.

1. Introduction

Healthcare as an innovation system differs quite distinctively from other sectors. First of all, public state health systems represent a major, in fact, in many cases *the* major, customers for innovations, but as customers they are extremely complicated and fragmented (Gabriel, Stanley & Saunders, 2017). Second, health innovation systems are unusual also from the perspective that the health systems as main innovation customers are different from the actual innovation users i.e., the patients and/or citizens (Gabriel, Stanley & Saunders, 2017). Innovations in healthcare relate not only to medical or pharmaceutical innovations (Ciani *et al.*, 2016), but also to new approaches to prevent illnesses, and promote the wellbeing of people (Gabriel, Stanley & Saunders, 2017). Digitalization, digital technologies and the use of heterogeneous data have had a major impact on opening innovation processes in the healthcare sector (Kalis, 2016). For instance, mobile device-assisted healthcare and medical applications are considered to create the next big advancement in the health industry (Balandin *et al.*, 2013; Francis Gomes & Moqaddamerad, 2016). Indeed, the definition of healthcare innovation has changed drastically over time: now innovations in patient care, wellness or health tech are considered as innovations in healthcare (Francis

Gomes & Moqaddamerad, 2016). The disruption that the use of data has caused in the healthcare sector, e.g., the availability of biomedical data and the genetic makeup, has even been compared to how the development of ICT changed the society in the past decades (Horgan *et al.*, 2014).

However, the reality is not as rosy as it may sound. Public health in general is affected by various kinds of determinants *outside* the healthcare system, such as economic, social, political and technological factors (Brownson *et al.*, 2009; Buse, Mays & Walt, 2012). Especially in preventive healthcare, which refers to use of these external determinants in health decision-making should be addressed to ensure robustness of public health policies. Preventive healthcare aims at anticipating future impacts and outcomes of the decisions made today (Sherrod, McKesson & Mumford, 2010) by incorporating proactive health and wellness service provisioning and early intervention, rather than treating symptoms and illnesses reactively. In other words, innovations in healthcare are also needed for bridging the gap between what is possible and the actual delivery of healthcare (Edenius *et al.*, 2010; Wass & Vimarlund, 2017).

One of the biggest challenges that public healthcare — and especially preventive healthcare — is suffering from is the lack of systematic distribution of knowledge flows, as characterized in the open innovation paradigm (Chesbrough, Vanhaverbeke & West, 2014). Despite increasing interest in exploring open innovation in the healthcare sector, there is a dearth of studies on open innovation in the public context (Wass & Vimarlund, 2016). The complexity of healthcare systems is one of the biggest challenges. Health policy making involves courses of action and inaction that impact different institutions, organizations, services as well as funding arrangements in the healthcare system in place (Buse, Mays & Walt, 2012; Dye, 2001). Healthcare is distinguished by the special status of biomedical knowledge in contemporary societies, as well as by experts who have mastered this knowledge (Gabriel, Stanley & Saunders, 2017). Moreover, open innovation in healthcare is challenged by "the things it has become famous for, including tight control of intellectual property rights and a certain amount of skepticism voiced by doctors and scientists who feel that their problems are so specialized that no one outside of their field could solve them" (Silvi, 2015).

Although healthcare data has been made increasingly available for decision-making, the lack of knowledge and adaptation of open innovation in public healthcare is manifested especially through the lack of *systematic* use of different types of data (Krumholtz, 2014), which has direct implications on health policies and policy making (Iivari *et al.*, 2017). In OECD countries, for instance, policy makers and health system managers seek to move toward performance-based governance, but this requires accurate and timely patient data, from actual care to health outcomes and costs (Paavola, 2017). However, at the moment, decision-makers, such as public health providers, community-level decision-makers, city-level decision-makers and governmental-level decision-makers, typically do not have any *control* over the design of the data, its formatting or how the data is collected. The opposite is in fact true; the knowledge health providers' base their decisions on is often severely fractured, disjointed, stored in multi-formats, and sometimes not even in an electronic format (Iivari *et al.*, 2017). Also, data-driven decision-making tools have been developed for health policy makers in Europe but even at their best, they rely on authenticated statistical data that is one to two years old — i.e., too old for preventive decision-making. Yet, this kind of historical data decision-makers ought to use to identify not only current needs but also to predict *future* needs and trends in specific thematic areas. Policy making in healthcare is often based on intuition rather than evidence and data (Otjacques, Hitzelberger & Feltz, 2014), because there is a lack of understanding on evidence-based policies (Brownson *et al.*, 2009). These conditions have resulted in significant challenges for preventive healthcare providers to use valid information for decision-making.

The healthcare sector has not widely engaged in open innovation (Wass & Wimarlund, 2016). Yet, there are demonstrable benefits of "a distributed innovation process based on purposively managed knowledge flows across organizational boundaries" (Chesbrough & Bogers, 2014, p. 17) also in the healthcare sector for health policy and public decision-making. Open innovation in data-driven policy making could indeed help to change the healthcare industry from treating sicknesses reactively into improving the wellness of people proactively (Clulow, 2013; Iivari *et al.*, 2017). The increase in the amount of and the diversity of information combined with improved storing capabilities for (electronic) data and analytical tools offer abundant opportunities to all stakeholders in the healthcare ecosystem (manufacturers, regulators, payers, healthcare providers, decision-makers, researchers) and moreover, data also enables improving general health outcomes when exploited the right way (Leyens *et al.*, 2017).

Accordingly, to address the gap on open innovation from the perspective of systematic distribution of knowledge flows, and data-driven decision-making, this chapter explores how preventive healthcare policy making could benefit from open innovation. In this vein, this chapter seeks to contribute to public policy discussion on open innovation, as well as provides empirically grounded research on how open innovation could advance data-driven preventive healthcare policy making?

The chapter is structured as follows. First, we discuss the literature on knowledge sharing and distribution in the healthcare sector, then we address the systemic features related to decision-making based on that knowledge, as well as address the specificities of data-driven decision-making. We will then present our methodological approach applied in the empirical case, present the key findings and conclude our discussion.

2. Knowledge-based Decision-Making in the Healthcare Sector

This chapter provides an overview on decision-making in the healthcare sector from the point of view of knowledge management.

2.1. *Knowledge in policy making*

Notions like knowledge sharing or knowledge management, which are closely linked with open innovation, are often applied either intentionally or unintentionally when it comes to policy making (Riege & Lindsay, 2006). Knowledge management as a process captures the collective expertize and intelligence internally and externally to an organization, and uses it to foster innovation through organizational learning (Yim *et al.*, 2004). The stronger the knowledge base, the more the policy decisions are supposed to succeed (Riege & Lindsay, 2006). "Knowledge" as a conceptual term features in the management literature as a strategic asset, whereas in the healthcare literature similar notions are often expressed with terms as "evidence" or "research" (Ferlie *et al.*, 2012). So, terms like knowledge, evidence and research are often being used interchangeably. For instance, Brownson *et al.* (2009) identified the missing element for public policy literature is a clear definition of evidence-based policy. Public policies refer to government policies or the policies of governmental agencies. Health policies concern courses of action and inaction which affect-involved institutions, organizations, services and funding arrangements of the healthcare system in place (Iivari *et al.*, 2017). Health policies in practice could incorporate both public and private policies. In addition, the development and deployment of health policies can occur in all levels of public decision-making, i.e., from national to regional to local levels. Policy makers play an important role in the process of innovation as they also intervene in various phases, with different kinds of consequences, e.g., in the market relationships between producers, innovators, users and patients (Ciani *et al.*, 2016).

While policy makers are under constant inquisition from society to improve effectiveness and quality of policy decisions despite limited resources (Keating & Weller, 2001), they are further expected to do so demonstrating better transparency and accountability (Riege & Lindsay, 2006). Riege and Lindsay (2006) assert that clear communication and partnership among involved stakeholders regarding the policy outcomes can be a starting point for better policy formulation through open knowledge management. Policy formulation in the context of healthcare is generally a complex process (Brownson *et al.*, 2009). One cause behind is that public health is influenced by numerous determinants outside the health system. When policy makers are formulating health policies, they also need to consider those external elements, such as scientific, economic, social and political forces (Brownson *et al.*, 2009; Buse, Mays & Walt, 2012). Therefore, public health policies have a great impact on the health status of populations in general. Otjacques, Hitzelberger and Feltz (2014) have claimed that slightly over half of public health policy makers are well informed by public health data before making decisions, and just half use data only sometimes or occasionally "never" for making public health policy decisions. Up to 64% of decision-makers never perform statistical analysis in making these decisions, and 57% of decision-makers do not use any simulation or forecasting and often only use census data and data from epidemiological studies (Otjacques, Hitzelberger & Feltz, 2014).

Van Beveren (2003) marks the necessity of stronger cooperation and communication among health entities for better patient-centered care. The public sector literature lacks discussion from a resource-based view (Ferlie *et al.*, 2012), where resources such as inter-entity-collaboration, open platforms, data, information, data integration, integration capacity, knowledge could be perceived as key policy enablers. Knowledge can be shared, sourced, discovered or created. Unfortunately, often health organizations are observed to obtain knowledge exclusively through acquisition (Van Beveren, 2003), while discovery or creation of knowledge within healthcare is a real possibility where ample amount of usable data is available.

2.2. *Data-driven policy making*

Public bodies are among the largest creators and collectors of data in many different domains (Janssen, Charalabidis & Zuiderwijk, 2012). The healthcare sector generally produces globally one of the highest amount of data in different forms (Raghupathi & Raghupathi, 2014), where the development of digital technologies, such as Internet of Things (IoT), and forthcoming 5th generation (5G) mobile networks, the prospects are immense for the use of data (Iivari *et al.*, 2017). Mostly discussed type of data are "big data", which was coined by Cox and Ellsworth (1997) who explained the visualizations of data, and challenges they posed for computer systems. As a concept, big data stimulates extremely and uncontrollably saturated digital contents that are used to generate information, in turn helping in knowledge creation (Lohr, 2012). However, data-driven policy making does not only rely on big data, but all types of heterogeneous "rich data" (Iivari *et al.*, 2017), meaning that data-driven policy making especially in preventive care is transitioning from volume toward several sources and formats of data. The various types of data utilized in policy making are essential in health systems especially when aiming for systemic innovations.

Discovery or creation of knowledge in big data, for example through data mining from huge volumes of data is conceptualized as "Knowledge Discovery in Databases" (KDD) (Fayyad, Piatetsky-Shapiro & Smyth, 1996). While in general, KDD can be applicable for one organization holding a big dataset themselves, this approach is also applicable for putting together rich and heterogeneous dataset sourcing from multiple stakeholders, as addressed in our chapter, to make sense of previously untapped knowledge sources. Rich and heterogeneous sources of data can offer significant opportunities for researchers, health professionals and policy makers to "move away from looking at population averages and toward the use of personalized information that has great potential to generate personal, societal and commercial benefits" (Heitmueller *et al.*, 2014). In the wave of digitalization, healthcare is transforming from a structured-based data (electronic patient report, diagnosis reports that are formally stored) toward semi-structured (home monitoring, telehealth, IoT devices, other sensor-based wireless devices) and unstructured (transcribed notes, paper prescriptions, discharge records, digital images, communication messages, radiograph films, MRI, CT images, ultrasound images, videos) forms of data (Raghupathi & Raghupathi, 2014; Wang, Kung & Byrd, 2018). Knowledge discovery based on data might be useful, but while big data is assumed to impact health sector positively, especially the inadequate integration of data in multiple healthcare information systems causes challenges (Wang, Kung & Byrd, 2018; Bodenheimer, 2005).

Therefore, in data-driven policy making, it is important for decision-makers to understand the different types of data sources that may be useful in healthcare-related policy making situations. The challenge for public policy making is that big data is mostly in the private sector. Three different data sources are typically used in the big data industries (Brownlow *et al.*, 2015). These are self-generated data, custom-provided data and free available data. The value of self-generated — i.e., personal data is growing (Schwartz, 2004). Personal data can generally refer to information generated by an individual, which is increasingly driving healthcare policies as well (Iivari *et al.*, 2017). Citizens want to receive more and more personalized and improved care based on their personal data, which impacts the healthcare system in a way that data and technologies allow patients to get care outside hospital walls to control and share their health information to other stakeholders in the healthcare ecosystem (Gabriel, Stanley & Saunders, 2017; Gaskell, 2017). Digitalization has therefore had a major impact on opening the innovation processes in the healthcare sector (Kalis, 2016). Governmental organizations typically collect personal data such as taxes, residence and date of birth; healthcare organizations maintain a variety of health records; businesses collect client data, shopping behavior, transactions, receipts, etc. (Ericsson, 2013). Free and/or open data in policy level decision-making (Janssen, Charalabidis & Zuiderwijk, 2012) could involve, e.g., traffic data, weather, geography, tourist information, statistics, business, public sector budgeting, and performance levels, policies and inspection (food, safety, education quality, etc.) (Janssen, Charalabidis & Zuiderwijk, 2012). Some examples show the open access of publicly funded data has offered great returns from the public investments providing policy makers data that is needed to address complex problems (Arzberger, Schroeder & Beaulieu, 2004). It has been claimed by Janssen, Charalabidis and Zuiderwijk (2012) that open data has no value in itself; it only becomes valuable when used. In this context however, little is known about the conversion of public data into services of public value.

2.3. *Systemic decision-making*

Decision-making environments are dynamic in the real world, and often tacit knowledge focused (Yim *et al.*, 2004). Knowledge-based decision-making that is supported by systemic thinking enables proactiveness of decision-making. When public health is concerned, it is important for policy makers to get accurate and real-time information in order to understand different dimensions of healthcare related problems, such as social care, and propose effective solutions for tackling them (Nieminen & Hyytinen, 2015). However, if we think the complexity of healthcare-related issues it is not even possible for one person to search and read enough information to guarantee robust information for decision-making. This is further challenged by the fact that even though healthcare data has been made increasingly available for decision-making, public healthcare is largely suffering from the lack of systematic use of different types of data (Krumholtz, 2014). Foresight is one approach among others that helps policy makers to strengthen the participatory, interactive and strategic elements of evaluation (Fetterman, 2001; Patton, 2011; Nieminen & Hyytinen, 2015). Foresight can be identified as "a systematic, participatory, future intelligence gathering and medium-to-long-term vision building process aimed at present-day decisions and mobilizing joint action" (Georghiou *et al.*, 2008, p. 11).

In order to improve the quality of care, efficiency and coordination, policy making and health system management is moving toward performance-based healthcare governance (Paavola, 2017). However, to enable performance-based governance, accurate and timely patient data is required, ranging from actual care to health outcomes and costs (Paavola, 2017). Often, health providers base their decisions on data knowledge that is fractured and fragmented and stored in various, e.g., in manual, formats (Iivari *et al.*, 2017). These conditions have resulted in significant challenges for preventive healthcare providers to use valid information for systemic decision-making. It is not of assistance to policy making that although there is increasing academic interest in exploring open innovation in the public healthcare sector that would involve system-level methods of collaboration, in order to make sense of the complexities of the healthcare system (Wass & Vimarlund, 2016). Health policy making involves courses of action and inaction that impact different institutions, organizations, services as well as funding arrangements in the healthcare system in place (Buse, Mays & Walt, 2012; Dye, 2001). Here, data would support the redesigning and evaluation of new models for healthcare service delivery, for instance, thus contributing to the discoveries and evaluations for new treatments. Encouraging the uptake of the most efficient and effective frameworks and practices to enable the collection, storage and use of personal health data to improve population health and to improve the effectiveness, safety and patient-centeredness of health care systems remains a significant policy challenge in many OECD countries" (Paavola, 2017).

As complex and dynamic systems (such as the healthcare system in our case) are constantly in a change and involve an enormous amount of impact indicators and within-system feedback loops, a dynamic approach is needed to able to address the complexity (Hargreaves & Podems, 2012). All approaches to strategic decision-making and management have both benefits and challenges. For instance, in many cases decision-making is still based on fragmented information which means that the comprehensive information on the environment and its change, as well as an understanding of wider short-term and long-term impacts, are often lacking in policy level decision-making (e.g., Loorbach & Rotmans, 2010). Therefore, open innovation through its key approach to knowledge distribution and interorganizational knowledge flows would have great impact on advancing the effectiveness and impact of policy making especially from a systemic perspective. Though driving toward increasingly porous innovation systems, knowledge, i.e., evidence and data should be generated more openly and in collaboration with external parties. Policy makers need to acknowledge that also healthcare-related data can come from anywhere, also through self-generated data by patients, needs of the patients and users should drive innovations in healthcare, coupled by the knowledge of practitioners

(Gabriel, Stanley & Saunders, 2017; Iivari *et al.*, 2017; Bullinger *et al.*, 2012). According to Gabriel, Stanley and Saunders (2017), open innovation initiatives in the healthcare sector would contribute to more efficient use of resources and enabling faster adoption and diffusion of healthcare innovations. Another objective is to contribute to innovation processes through deeper understanding of health systems as well as the needs of patients or citizens. In political terms, open innovation should make health innovations more democratic through demand-driven approaches.

3. Research Design

The empirical investigation in this study was collected within a European Commission Horizon 2020 funded project on the use of meaningful data for healthcare policy making. The empirical study is based on the qualitative, semi-structured expert interviews conducted with regional public bodies and city-level healthcare policy makers in the province of Northern Ostrobothnia, Finland, during Spring 2017. The focus of the interviews was to investigate the practices related to data-driven knowledge management, and specifically what implications different types of data, information and knowledge sources have on policy making in preventive healthcare. The interviews provided a preliminary understanding of the phenomenon in question, and what kinds of needs, barriers and future opportunities relate to data-driven decision-making. The specific case identified for preventive healthcare was the mental health of young people.

Eight representatives from different city and municipal organizations were interviewed in six different interview sessions, either alone or via group interviews (Table 1). The interviewed were conducted in the native language of the interviewees, i.e., in Finnish. All interviews were recorded, transcribed and translated into English, and thematically analyzed (Boyatzis, 1998).

In order to further advance knowledge distribution and decision-making in preventive healthcare, a group modeling workshop was organized on 3 May 2017, where all interviewed parties were invited to collaboratively validate, visualize and simulate the interview

Table 1. Interviewed policy makers

Role	Date (2017)	Duration (h:mm)
Finance Manager, Development and Quality Manager	Feb 27th	1:02
Director of Healthcare and Social Welfare	Mar 20th	0:53
Head of Health, Social and Education Services	Mar 22nd	1:24
Director of Healthcare, Director of Social Welfare	Mar 22nd	0:59
Director of Education	Mar 28th	1:06
Director of Joint Municipal Authority	Apr 10th	1:33

findings with the help of system dynamic modeling techniques. System dynamic modeling is a method that combines both tacit and deliberate knowledge of individuals and subgroups within organizations, where alternative actions are used to simulate and predict future outcomes (Woodside, 2010). During a system dynamic modeling workshop, the complex causal dependencies of various concepts are identified and visualized into a dynamic hypothesis, the structure of which is validated through interactions between workshop participants from various stakeholder groups.

A system dynamic (SD) model describes complex connections between multiple elements in different levels, in addition, it explains dynamic processes with feedback in the system (He *et al.*, 2006). For policy analysis and recommendation, SD models help predict complex system behavior under various "what-if" scenarios (Mohapatra, Mandal & Bora, 1994). SD modeling techniques challenge narrow views and encourage seeing the big picture both in time and space, that is, considering outcomes both in the short and long run and across organizational boundaries. The models foster communication between different views enabling reflective and collaborative solutions. All of these features of SD modeling are building blocks of innovations and eventually lead to better decisions.

More specifically, we applied group model building (Vennix, 1996; Michaud, 2013) with focus on qualitative modeling as the tool to facilitate discussion and sharing of knowledge and understanding among the different decision-makers. This tool was chosen, as it allows slicing and presenting complex systems on a suitable level of abstraction in order to identify the points of interconnections between knowledge flows that otherwise could fall in different silos in organizational boundaries, both internally and externally to healthcare organizations. Moreover, group model building stands apart from other modeling techniques by the direct involvement of decision-makers, stakeholders and topical experts. It has two primary aims: elicit input from the participants to construct and validate a model, and use the modeling process as a learning process for the participants as they share knowledge with each other and reflect their mental models with the model under construction. The two aims share the underlying objective of aiding decision-making. We employed group model building to facilitate participation of a heterogeneous group of actors involved in various areas in preventive healthcare, including those of the social welfare, healthcare, education and other sectors, i.e., service units. Through a set of methods and work practices, we elicited information and insight needed for system dynamics models, while helping decision-making as a process on its own.

Therefore, we see that system dynamic modeling is one of the key tools in driving open innovation in the healthcare sector, providing evidence-based support for policy makers to promote the purposively managed, distributed knowledge flows across the silos of public decision-making.

4. Results

The research results were categorized under two themes. The first includes current decision-making and data-driven knowledge management practices in relation to preventive healthcare, and mental health of young people. The second one presents the knowledge requirements and needs that policy makers have for producing the healthcare policies of the future. These themes emerged as the most important factors in discovering the role of open innovation in data-driven healthcare policy making.

4.1. *Preventive decision-making*

Organizing resources for preventive mental healthcare services for young people was highlighted as the most crucial challenge for decision-makers. In this context, substance abuse and unemployment prevention emerged as very close concerns for decision-makers. Thus, mental health is a complex and multi-leveled case, and data required for policy making requires the use, analysis and visualization of heterogeneous types of data.

However, the research results reveal that as a starting point, there was high variation in the way that the interviewed policy makers were using digital systems in current or historical evaluation of healthcare situations and cases. Some of the interviewed policy makers did not have any digital tools in use in the decision-making process. In fact, for some of them, a typical situation is that the policy makers are having secretaries who are preparing the information for them based on the information that can be found from internet and statistical data.

> ...most often we use the information on what has happened, not information what is most likely going to happen.
>
> Real-time data? Capacity, situation, daily, customer flows, use of spaces etc. Useful for the organiser who is responsible for the entity. Production unit, individual follow-up. Intensive care, customer-centric impact. Long term data...
>
> The real-timeliness of data matters, in the wellness report there is a lack, since it's updated every two years. We would need information from the on-going year... Something could come faster... Some have data on booklets, some have manual data... How does information travel between different actors, do we buy child protection and substance services and elderly services. Could be even faster, a substance client needs information right away.

The systemic evaluation based on interorganizational knowledge sharing, including qualitative and quantitative approaches, was seen as an approach that would provide the needed information about the past and current state of the system and the health status of individuals, which was referred to as the digital footprint. However, this is not yet the case in current decision-making models. In our case, system dynamic modeling provided the needed evidence of dependencies that follow from actions taken. The systemic evaluation will also support redirecting policy instruments to better respond to the needs of a shaping healthcare environment, and to show the overall results and outcomes of potential decisions, in order to support more preventive and predictive decision-making.

An interesting notion from the interviews was also the problem with scarce data from social services compared to data from health services. Legislation is one key challenge there, as of now, in many situations, laws prevent the personal identification of individuals, as well as the use of data across organizational boundaries, such as from education or social services into the use of healthcare decision-makers. Moreover, healthcare data mainly consists of numerical, quantitative data and qualitative data that would be most relevant to preventive actions are lacking.

> We have customer data missing in the social services side, quite a lot actually. Like from private clients. Kids, youngsters, families with children, working people... We have quite a lot of data related to use, but no predictive information really.

The interviewees pointed out that this issue needs urgent development, as preventive mental healthcare services are a much wider concern for the society than say, for example, diabetes. It was said, that if a person has diabetes, that concerns just the individual, but if a person has mental health issues, that concerns also the immediate family, relatives, school, work and a person's ability to contribute to society and economy, e.g., in the form of employment and taxation.

4.2. *Knowledge distribution in decision-making*

Separating knowledge distribution and knowledge management from preventive decision-making were seen as two sides of the same coin. However, based on the use of data, some key issues were identified.

> The problem is that cause-consequence isn't that clear. Education and culture plus wellness services, why does it show that the red services have increased? There has also been a lot of leukaemia and premature birth cases. If we rely too much on raw data, we can draw false conclusions.
>
> Integrating research results would be useful. We would need more, like impact evaluations.

One challenge of the evaluation leading to actual decision-making is also analyzing the interdependencies and interactions within the healthcare in connection to social care system, as well as between the system and environment. In our case it was also noted:

> The information is shared in silos and it is owned by different actors. It would be great to understand for instance what services has been offered to one individual. Perhaps a person has got a disability diagnosis later. Perhaps he has been going through the whole school system and his situation is never investigated.
>
> Looks like that with youngsters, we have this group that doesn't put effort into studying academic subjects, there should be more of these apprenticeship type education for those who don't read. Those that have been under special needs education, have received better scores, but they fall out when they no longer get support. Those that are worse than normal students, don't get into schools... There could be developmental delays, behavioural problems, substances, child protection". This information is divided and behind different authorities. What services have been offered to one person, for example ... Has this person gotten a developmental diagnosis later. Have they just surfed through school so that they have never been examined? Now they have been guided to employment services and never received (healthcare) services that they would have been entitled to. These kinds of things should be summed up.

Among the policy makers that we interviewed, foresight was seen as a way to generate information about alternative futures in a system by analyzing trends and drivers that cause changes in the system.

> Scenarios would be interesting. If people for instance loose some reimbursement or state aid, it might have some long term social impact that we do not see now.

Another key source of external knowledge was seen to emerge from the private sector, such as from schools, church, sport centers, clubs, libraries, youth centers, private sports facilities, cultural facilities and so on. The way public decision-makers are collaborating with the private sector and third parties is varying much between the regions, but all of the interviewees acknowledge that important information could be gained from membership registries and especially from schools. However, access to private sector data is an issue, as so far, as here as well, the legislation prevents the identification of an individual. However, it was stated that for preventive care, that is exactly the need, in order to capture those youngsters before they develop serious mental conditions, substance abuse or issues with abuse, crime and so on. Active collaboration with schools was highlighted and schools also have access to collaboration with various kinds of third parties.

Not only access and sharing of knowledge, whether public or private, was an issue but also the depth of it:

> The information is too general level at the minute... with the current data we can make false conclusions.

Evaluation approaches will be in the future combined with foresight to generate detailed information on the development of a phenomenon at the systemic level. This approach will support operational target setting by providing information of the potential impacts of planned actions related to decision, and how the policies could impact people's mental health or well-being in the near and longer future.

Foresight is also addressing the challenge of making sense of rapidly changing decision-making contexts and aiding formulation of commonly shared future visions among central actors in the system. This is particularly important in the case of mental health in which it is crucial to identify the risk indicators and to react early in order to help individuals when things can possibly be affected more easily and with smaller costs:

> We would need the preventive and predictive system that helps us to see the future.

To synthesize the findings from the interviews and the system dynamic modeling workshop, the key factors in data-driven healthcare policy making are illustrated in Fig. 1.

The key factors are increasingly open and more widely distributed and utilized healthcare data, coupled with increasingly future-oriented, predictive and causation-supported data as the basis of knowledge-based decision-making. Preventive stance to decision-making is not possible without access to various types of data, but timeliness and accuracy of that data is equally relevant. The reliability of data for policy making can be supported through accessing data from various sources beyond single organizational domains, as mental health as the case especially requires as open and as distributed flows of knowledge, in order to produce as preventive and as tailored policies and services as possible.

5. Discussion and Conclusions

Through exploring the utilization of different kinds of data in policy making, this study highlights the complexity of decision-making in the healthcare sector (Gabriel, Stanley & Saunders, 2017; Wass & Vimarlund, 2016). Focusing on a single type of data from a single sector cannot truly uncover the systemic nature of relations and dependencies, which not only policy makers but healthcare practitioners alike need to acknowledge.

Here, the technological and analytical solutions brought by digitalization have the ability to support faster and better use of

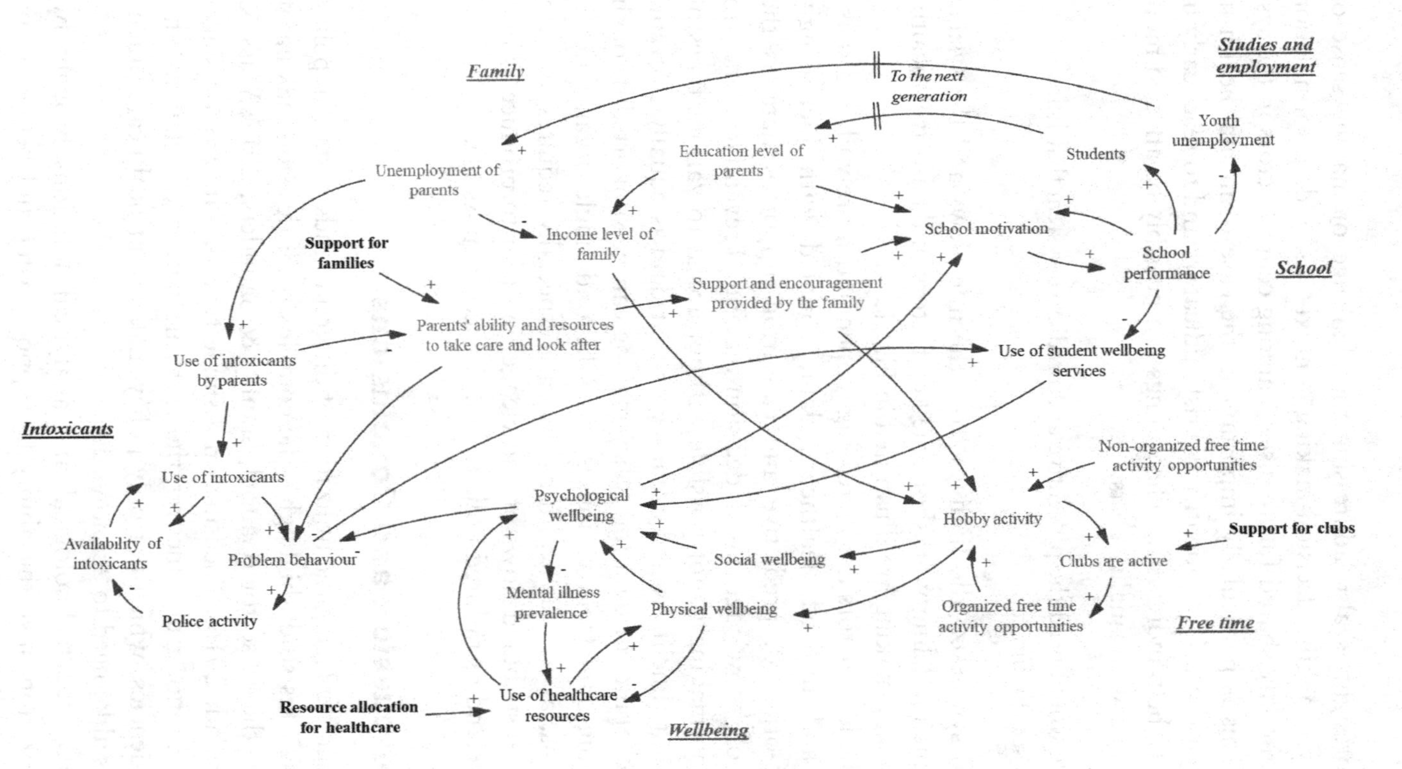

Fig. 1. The development of data-driven decision-making

data for creating more personalized and tailored services for the needs of individuals and their families. However, they are effective in preventive healthcare only if knowledge is systematically carried across organizational boundaries. Enabling and supporting distributed knowledge flows for policy decisions is only one preliminary step in the road to data-driven policy making. Data analysis and visualization are essential elements in turning data into information for decision-makers. Moreover, as traditional data concentrates on current and historical statistics whereas decision-makers increasingly require alternative futures and long term impacts of the decisions made, there is a definite need for further research in integrating rich data to other tools for healthcare policy development. Decision-making in healthcare sector has direct and indirect implications on individual patients, health professionals, health businesses as well as the society as a whole. Our results show how open innovation could advance this further, as currently, decision-making still mainly occurs in individual silos, and knowledge does not travel across organizational and departmental boundaries. Openness in the distribution of knowledge at the moment makes preventive decision-making challenging, as decision-makers do not have access to systemic level, analyzed data, or the data becomes available only after a delay due to processing by other official statistics collecting organizations. Therefore, data-driven decision-making is always slightly retrospective, especially when more and more data sources are identified and utilized for drafting specific decisions or policies. Only through reigning the tacit knowledge with the support of suitable methods, such as system dynamics, and the right types of data, it can lead to better preventive decision-making.

This study provides empirically grounded findings on how different types of data and knowledge sources are and should be distributed across organizational boundaries. By looking at the preventive mental healthcare of young people, we explored what type of knowledge is distributed across city service units and what implications these data-driven decisions have had on healthcare policy at the systemic level. Thus, this study contributes to open innovation in the public context and open innovation in healthcare. We seek to contribute to the discussions on how open innovation paradigm could advance the development of (better) data-driven policies, and support knowledge management and decision-making in public organizations. We also contribute to data-driven decision-making. Alike, this study also addresses issues in relation to innovation systems literature by addressing the healthcare system.

The main practical implications of the study relate to the opportunity to increase knowledge on the applicability of open innovation in the healthcare sector, and how this can be advanced in practice through systems thinking, and system dynamic modeling as a methodological approach. Thus, the ways in which open innovation as a process of distributed knowledge flows across organizational boundaries can increase the effectiveness of preventive healthcare that can be examined through participatory group model building, which was applied in this study. In this way, this study also highlights that the greatest barriers to advancing open innovation in healthcare relate to the distribution of knowledge rather than the availability of healthcare data as such. Without breaking the silos and allowing knowledge to flow between different service units internally to public healthcare organizations, but also between public and private sector, preventive actions cannot have the strongest impact on the health of the society and especially our youth. Here, public policy makers have a key role in driving this change and opening the healthcare sector further.

The system dynamic group building proved to have a multifaceted effect in advancing open innovation. Firstly, the process widened the scope of the mental models of different stakeholders to understand the big picture and the ripple effects of decisions at one sector to another (such as, decisions concerning education and social services affecting healthcare sector outcomes). Second, it distilled the essential data needs for assessing the impacts of systemic innovations, i.e., guides the data collection and fusion process. Third, it showed the unwanted side-effects, counter-intuitive behavior of complex systems and policy resistance structures, the knowledge of which can be used for overcoming the implementation challenges.

Although the research methodology utilized in this study is its strength, it is also the weakness. Through such a strong qualitative stance, national system-level correlations and statistical cause–consequence analyses cannot be made. Although the phenomenon itself, preventive healthcare, especially in the case of the youth, is something that touches most modern, advanced societies, the data in this study includes also the largely experimental, tacit type of knowledge, which is difficult to quantify.

However, these challenges also lead to interesting future research directions. Methodologically, a longitudinal approach on following how policy making is changing with the use of more predictive types of data, supported by systemic, collaborative decision-making methods, would allow a more systematic analysis on how open innovation paradigm is spreading in the healthcare sector, especially among public healthcare providers, not just in private corporations. Moreover, it would be interesting to study how the mindsets of policy makers are changing with the use of open innovation practices and system dynamic tools, and how public healthcare providers embrace openness in order to advance the well-being of citizens as a whole.

To summarize the study on how open innovation could advance data-driven preventive healthcare policy making, we see it as two-fold process: first, increasing understanding on the importance of freer flows of knowledge across organizational boundaries for better healthcare is the preliminary step. Second, with appropriate tools, such as system dynamics, we are able to concretize this knowledge into actionable, usable relations and correlations. Only with proper understanding coupled with right tools, policy makers are truly able to utilize the right types of data in preventive decision-making.

Acknowledgement

This research was funded by The European Union H2020-SC1-2016-CNECT SC1-PM-18-2016 — Big Data Supporting Public Health Policies in the project called Meaningful Integration of Data, Analytics and Services (MIDAS), Grant Agreement No. 727721. The authors would like to acknowledge the MIDAS project consortium for their support.

References

Arzberger, P., Schroeder, P. & Beaulieu, A. (2004). An international framework to promote access to data. *Science, 303*(5665), 1777–1778.

Balandin, S., Balandina, E., Koucheryavy, Y., Kramar, V. & Medvedev, O. (2013). Main trends in mHealth use scenarios. *Journal on Selected Topics in Nano Electronics and Computing, 1*(1), 64–70.

Boyatzis, G. E. (1998). *Transforming Qualitative Information: Thematic Analysis and Code Development*. Thousand Oaks, CA: Sage Publications.

Bodenheimer, T. (2005). High and rising health care costs. Part 2: Technologic innovation. *Annals of Internal Medicine, 142*(11), 932–937.

Brownlow, J., Zaki, M., Neely, A. & Urmetzer, F. (2015). Data and analytics — Data-driven business models: A blueprint for innovation. Working Paper, 2015.

Bullinger, A. C., Rass, M., Adamczyk, S. & Moeslein, K. M. (2012). Open innovation in health care: Analysis of an open health platform. *Health Policy, 105*, 165–175.

Buse, K., Mays, N. & Walt, G. (2012). *Making Health Policy*. UK: McGraw-Hill Education.

Chesbrough, H. & Bogers, M. (2014). Explicating open innovation: Clarifying an emerging paradigm for understanding innovation. In *New Frontiers in Open Innovation*, eds. Chesbrough, H., Vanhaverbeke, W. & West, J., Oxford: Oxford University Press.

Chesbrough, H., Vanhaverbeke, W. & West, J. (eds.) (2014). *New Frontiers in Open Innovation*. Oxford: Oxford University Press.

Ciani, O., Armeni, P., Boscolo, P. R., Cavazza, M., Jommi, C. & Tarricone, R. (2016). De innovatione: The concept of innovation for medical technologies and its implications for healthcare policy-making. *Health Policy and Technology, 5*, 47–64.

Clulow, S. (2013). Open innovation strategies in the healthcare industry. NineSigma Whitepaper.

Cox, M. & Ellsworth, D. (1997). Application-controlled demand paging for out-of-core visualization. In *Proceedings of the 8th conference on Visualization' 97*, Phoenix, AZ, 18–24 October 1997, IEEE Computer Society Press, Los Alamitos, CA.

Dye, T. R. (2001). *Top Down Policymaking*. Chatham House Pub.

Ericsson (2013). *Personal information economy — Consumers and the evolution of commercial relationships*. An Ericsson consumer insight summary report. Retrieved from https://www.ericsson.com/res/docs/2013/consumerlab/personal-information-economy.pdf.

Fayyad, U., Piatetsky-Shapiro, G. & Smyth, P. (1996). From data mining to knowledge discovery in databases. *AI magazine*, *17*(3), 37–54.

Ferlie, E., Crilly, T., Jashapara, A. & Peckham, A. (2012). Knowledge mobilisation in healthcare: A critical review of health sector and generic management literature. *Social Science & Medicine*, *74*(8), 1297–1304.

Fetterman, D. (2001). The transformation of evaluation into a collaboration: A vision of evaluation in the 21st century. *American Journal of Evaluation*, *22*(3), 381–385.

Francis Gomes, J. & Moqaddamerad, S. (2016). Futures business models for an IoT-enabled healthcare sector: A causal layered analysis perspective. *Journal of Business Models*, *4*(2), 60–80.

Gabriel, M., Stanley, I. & Saunders, T. (2017). Open innovation in health: A guide to transforming healthcare through collaboration. May 2017, Nesta, London.

Gaskell, A. (2017). *The Collaborative Nature of Healthcare Innovation.* Forbes. 19 June 2017. Retrieved from https://www.forbes.com/sites/adigaskell/2017/06/19/the-collaborative-nature-of-healthcare-innovation/#4ac29a8179ad.

Georghiou, L., *et al.* (2008). *The Handbook of Technology Foresight.* Cheltenham: Edward Elgar Publishing.

Hargreaves, M. B. & Podems, D. (2012). Advancing systems thinking in evaluation: A review of four publications. *American Journal of Evaluation*, *33*, 462–470.

He, C., Okada, N., Zhang, Q., Shi, P. & Zhang, J. (2006). Modeling urban expansion scenarios by coupling cellular automata model and system dynamic model in Beijing, China. *Applied Geography*, *26*(3), 323–345.

Heitmueller, A., Henderson, S., Warburton, W., Elmagarmid, A., Pentland, A. & Darzi, A. (2014). Developing public policy to advance the use of big data in health care. *Health Affairs*, *33*(9), 1523–1530.

Horgan, D., Romao, M., Torbett, R. & Brand, A. (2014). European data-driven economy: A lighthouse initiative on Personalised Medicine. *Health Policy and Technology*, *3*, 226–233.

Iivari, M., Francis Gomes, J., Pikkarainen, M., Häikiö, J. & Ylén, P. (2017). Digitalisation of healthcare: Use of data in policy making. In *Proceedings of the XXVIII ISPIM Innovation Conference*, 18–21 June, Austria, Vienna.

Janssen, M., Charalabidis, Y. & Zuiderwijk, A. (2012). Benefits, adoption barriers and myths of open data and open government. *Information Systems Management*, *29*(4), 258–268, DOI: 10.1080/10580530.2012.716740.

Kalis, B. (2016). The time is now for open innovation in healthcare. Insight driven health blog. Accenture. 30 December 2016. Retrieved from https://www.accenture.com/us-en/blogs/blogs-time-now-open-innovation-healthcare.

Keating, M. & Weller, P. M. (2001). Rethinking governments roles and operations. In *Are You Being Served? State, Citizens and Governance.* Allen & Unwin.

Krumholtz, H. M. (2014). Big data and new knowledge in medicine: The thinking, training, and tools needed for a learning health system. *Health Affairs*, *33*(7), 1163–1170.

Leyens, L., Reumann, M., Malats, N. & Brand, A. (2017). Use of big data for drug development and for public and personal health and care. *Genetic Epidemiology*, *41*, 51–60.

Lohr, S. (2012). The age of big data. *New York Times*, 11 February. Retrieved from <http://www.nytimes.com/2012/02/12/sunday-review/big-datas-impact-in-the-world.html>.

Loorbach, D. & Rotmans, J. (2010). The practice of transition management: Examples and lessons from four distinct cases. *Futures*, *42*, 237–246.

Michaud, W. R. (2013). Evaluating the outcomes of collaborative modeling for decision support. *Journal of the American Water Resource Association*, *49*(3), 693–699.

Mohapatra, P. K., Mandal, P. & Bora, M. C. (1994). *Introduction to System Dynamics Modeling*. University of Nevada Press.

Nieminen, M. & Hyytinen, K. (2015). Future-oriented impact assessment: Supporting strategic decision-making in complex socio-technical environments. *Evaluation*, *21*(4), 448–461.

Otjacques, B., Hitzelberger, P. & Feltz, F. (2014). Interoperability of e-government information systems: issues of identification and data sharing. *Journal of Management Information Systems*, *23*(4), 29–51.

Paavola, H. (2017). Towards open health innovation — openness of research, development and innovation activity in health sector in Finland. Final report, 5 April 2017. Tempo Economics Oy.

Patton, M. Q. (2011). *Developmental Evaluation: Applying Complexity Concepts to Enhance Innovation and Use*. New York, NY: Guilford Press.

Raghupathi, W. & Raghupathi, V. (2014). Big data analytics in healthcare: Promise and potential. *Health Information Science and Systems*, 2(3). Retrieved from https://doi.org/10.1186/2047-2501-2-3.

Riege, A. & Lindsay, N. (2006). Knowledge management in the public sector: Stakeholder partnerships in the public policy development. *Journal of Knowledge Management*, *10*(3), 24–39.

Schwartz, P. M. (2004). Property, privacy and personal data. *Harvard Law Review 2055*, *117*(7). Retrieved from http://scholarship.law.berkeley.edu/facpubs/2150.

Sherrod, D., McKesson, T. & Mumford, M. (2010). Are you prepared for data-driven decision making? *Nursing Management*, *41*(5), 51–54.

Silvi, J. (2015). Here, there, everywhere — The rise of open innovation in healthcare. *GE LookAhead*, 13 February 2015.

Van Beveren, J. (2003). Does health care for knowledge management? *Journal of Knowledge Management*, 7(1), 90–95.

Vennix, J. (1996). *Group Model Building*. New York, NY: Wiley.

Wang, Y., Kung, L. & Byrd, T. A. (2018). Big data analytics: Understanding its capabilities and potential benefits for healthcare organizations. *Technological Forecasting and Social Change*, *126*, 3–13.

Wass, S. & Vimarlund, V. (2016). Healthcare in the age of open innovation — A literature review. *Health Information Management Journal*, *45*(3), 121–133.

Woodside, A. G. (2010). Bridging the chasm between survey and case study research: Research methods for achieving generalization, accuracy, and complexity. *Industrial Marketing Management*, *39*, 64–75.

Yim, N., Soung-Hie, K., Hee-Woong, K. & Kee-Yong, K. (2004). Knowledge based decision making on higher level strategic concerns: System dynamics approach. *Expert Systems with Applications*, *27*, 143–158.

Chapter 8

Early Experience with Open Data from CERN's Large Hadron Collider

Achintya Rao[¶], *Sünje Dallmeier-Tiessen*[†], *Kati Lassila-Perini*[§], *Thomas McCauley*[*] *and Tibor Šimko*[‡]

[*]*University of Notre Dame, USA*

[†]*Scientific Information Service, CERN, Switzerland*

[‡]*Department of Information Technology, CERN, Switzerland*

[§]*Helsinki Institute of Physics, Finland*

[¶]*University of the West of England, Bristol, UK and CMS Experiment, CERN, Switzerland*

This chapter covers perspectives from the various partners who have worked on the release of large volumes of open research data from the Large Hadron Collider via the CERN Open Data Portal. The early experiences mentioned in the title refer to the launch of the Portal in November 2014 with the release of the first batch of high-level research data collected in 2010 by the CMS Collaboration. This chapter covers the motivation for releasing particle physics data openly as well as the challenges faced in doing so and solutions developed to facilitate these efforts. The authors also touch upon the use cases of the open datasets and the impact the first release has had. Caveat lector: *The experiences and figures described in this chapter correspond to the year 2016 when this piece was originally written. The reader may want to consult Nature Physics 15 (2019) 113–119 to learn about later developments.*

1. Introduction

Located on the outskirts of Geneva on both sides of the Swiss-French border, CERN, the European Laboratory for Particle Physics, is the world's premier research facility for accelerator-based high-energy physics. The laboratory's flagship accelerator is the Large Hadron Collider (LHC), which, at 27 kilometres in circumference, is the biggest and most powerful particle accelerator ever built. It accelerates protons in clockwise and anti-clockwise directions to nearly the speed of light before colliding them at four points on its ring. Gigantic particle detectors, operated by international collaborations of scientists and engineers, are located at each of these collision points. They record information from these collisions, generating an enormous volume of data for analysis: each collision event produces data of the order of megabytes (MB), and, with the LHC delivering collision events 40 million times each second, the CERN Data Centre has collected tens of petabytes (PB) since the accelerator began operations in 2010.

The collisions produced by the LHC allow physicists to study the fundamental particles and forces of nature, and test the various theories and models that have been proposed to explain their behavior. Members of the LHC collaborations perform this research by analyzing the collision data and share the results in open-access publications. These datasets are truly unique in many ways, and, outside of the particle physics community, not only are they of interest to students and data scientists, they also form an important part of the scientific legacy of the LHC.

Now, a large portion of these data are being made publicly accessible without any restrictions to anyone in the world with an internet connection. In 2014, the CMS Collaboration released 27 terabytes (TB) of data that were recorded in the second half of 2010, representing half that year's data harvest.

Following this successful release, and equipped with the lessons learnt from the experience, CMS released a second batch of data in April 2016[1] (300 TB in total, including 100 TB of collision data recorded in 2011). However, this chapter focuses on the first release from November 2014.

2. Motivation

The release of open data from the LHC is motivated, in part, by the desire to ensure that these data continue to remain available to researchers in the future. The four LHC collaborations have therefore adopted policies for data preservation and access.[2] These policies include related matters such as embargo periods, licensing and reuse. The specifications in the data preservation policies distinguish four different data levels that help define the resulting recommendations[3]:

1. Level 1 data comprises data that is directly related to publications which provide documentation for the published results;
2. Level 2 data includes simplified data formats for analysis in outreach and training exercises;
3. Level 3 data comprises reconstructed data and simulations as well as the analysis-level software to allow a full scientific analysis;
4. Level 4 covers basic raw-level data (if not yet covered as Level 3 data) and their associated software and allows access to the full potential of the experimental data.

The open data being discussed in this chapter refer to Levels 2 and 3.

[1] http://cms.web.cern.ch/news/cms-releases-new-batch-research-data-lhc.
[2] http://opendata.cern.ch/collection/Data-Policies.
[3] See https://arxiv.org/abs/0912.0255.

CERN is a strong proponent of openness in research, having worked with publishers and other laboratories to ensure that all particle physics results from the LHC collaborations are published as open-access papers.[4] More recently this paradigm has been expanded to cover Open Science more comprehensively.[5] CERN has helped build tools and digital libraries to foster Open Science practices beyond the particle physics community, such as the widely used Invenio digital library software. Invenio forms the basis for projects such as ZENODO, which has helped thousands of scientists archive their work on CERN's servers with persistent identifiers. In order to facilitate storage for and access to open data from the LHC collaborations, the laboratory launched the CERN Open Data Portal, built using Invenio, in November 2014.

All four LHC experiments provide open data in a format suitable for a classroom environment, typically used in the Physics Masterclasses[6] for high school students around the globe (see also Section 5.1). In additional to these "educational" datasets, the CMS Collaboration, in accordance with its data preservation and data access policies, has also released large volumes of high-quality data for research purposes along with the tools necessary for accessing and analysing them. This chapter describes the motivation behind the launch of the CERN Open Data Portal as well as the background to and experience with the first release of open data from the LHC.

3. Challenges

3.1. *Volume of data*

Making data public does not make them any simpler. Since the LHC began delivering collisions in 2010, and as of writing this document (June 2016), the four detectors have recorded several tens of PB of collision data. Performing an analysis on this volume of data involves

---

[4] See https://scoap3.org/ and https://cds.cern.ch/record/1955574.

[5] http://home.cern/cern-people/opinion/2014/11/road-open-science.

[6] http://physicsmasterclasses.org/.

several stages. First, the data from the collisions, which are in the form of electrical signals from individual channels within each detector, must be pieced together, or reconstructed, to form a coherent picture of the collision itself. Then, the analysis software must sift through the collisions to look for signatures associated with the predicted physics phenomena or for deviations from the predictions.

The LHC collaborations rely on the Worldwide LHC Computing Grid (or the Grid, for short)[7] to perform these analysis tasks. The Grid is a network of interconnected computers distributed all over the globe, and datasets are broken into chunks to facilitate processing. While those outside the LHC collaborations do not typically have access to these resources, it is worth pointing out that the research-level open data from the detectors are already "reconstructed" and "reprocessed" (Level 3 data) — that is, a lot of the gruntwork has already been done to convert the datasets into a form useful for the final analyses themselves (see also Section 6.2). Despite this, performing research with the "analysable" data is no small task and preparing an appropriate dataset, depending on the analysis one wishes to perform, can take up considerable time and computing resources. To make life easier, the Portal provides smaller, simplified datasets — known as "derived datasets" — that can be used for analysis in a classroom or university environment without a large technological overhead. The Portal also makes available the code used to prepare these derived datasets.

3.2. *Software environment and documentation*

Even in a format that includes meaningful objects reconstructed from raw data, the data from the LHC still remain very complex. The data files cannot be opened and understood as simple data tables — they contain special structures with all the necessary information for data interpretation — and dedicated software is necessary to analyse them. The software and an appropriate working environment must be provided together with the open data. Using these data for scientific purposes requires good knowledge of experimental

[7] http://wlcg.web.cern.ch/.

particle physics and functioning of a detector, and documentation can hardly cover all these aspects, understanding of which is usually acquired during undergraduate and graduate studies.

It is also important to remember that the release of the data happens some years after these data were collected. Roles within the collaboration change regularly and those responsible for various data-collection efforts at the time of data-taking are not necessarily involved in those activities come data release. Therefore, significant efforts are needed to acquire and capture the relevant documentation and preserve the institutional knowledge of the collaboration.

4. Solution

A collaborative approach was taken to deliver both the CERN Open Data Portal and the first big open data release — from CMS — at the same time. Members of CERN IT, CERN Scientific Information Service and the CMS Collaboration developed the platform, catalogued the data and analysis tools for the release, and prepared the appropriate user guidance and documentation. From the very beginning, it proved to be important to have subject specialists, information specialists and IT experts on board together to tackle the delivery of the big, complex datasets. Regular working meetings, coding/documentation sprints and feedback cycles ensured that the work on the Portal and its content was on track to meet the requirements of the potential users and use cases.

4.1. *Developing the platform*

The CERN Open Data platform is based on the Invenio digital library framework,[8] which was originally developed at CERN and is now co-developed by an international collaboration.

Invenio provides an ecosystem of standalone independent packages that can be used to build custom digital repositories. Use cases for Invenio include integrated library systems, digital document

[8] http://inveniosoftware.org/.

repositories, multimedia archives and data repositories. The software framework has been used for numerous such instances around the world and is particularly known for its scalability for large libraries, such as CERN Document Server (CDS)[9] or INSPIRE.[10]

Invenio is mainly used in academic environments. In order to make the CERN Open Data Portal attractive to the general public, including high school students and other targeted audiences, several features such as documentation pages and glossaries have been included on the Portal. Figure 1 shows the home page of the CERN Open Data Portal.

A crucial aspect of providing open data is to ensure the long-term preservation of the datasets, as well as the accompanying software tools. Invenio enables the persistence and longevity of the Portal's assets, drawing inspiration from recommended practices of the Open Archival Information System (OAIS).[11] As part of this,

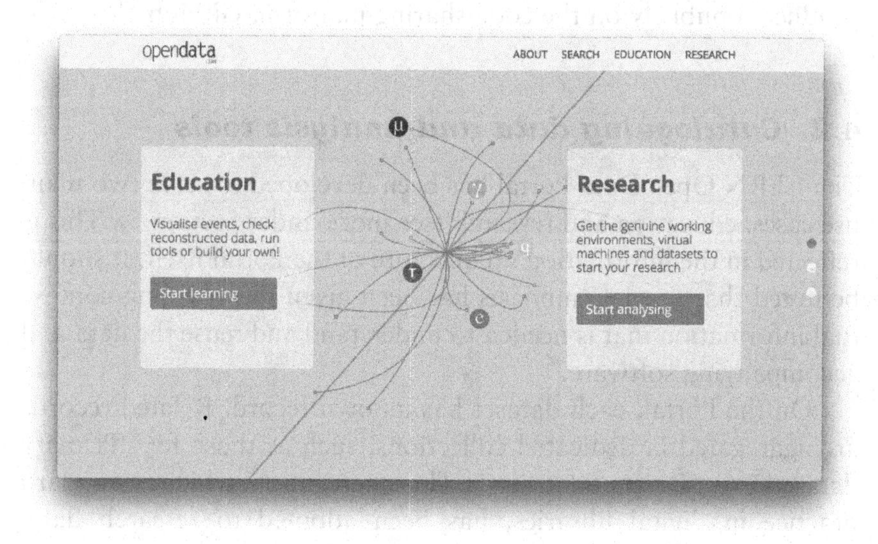

Fig. 1. CERN Open Data Portal home page

[9] https://cds.cern.ch/.
[10] https://inspirehep.net/.
[11] https://en.wikipedia.org/wiki/Open_Archival_Information_System.

persistent identifiers — in this case Digital Object Identifiers (DOIs) — ensure that the datasets and software tools can be referenced and cited well into the foreseeable future.

The datasets available on the CERN Open Data Portal are themselves stored in CERN's EOS[12] disk storage system. EOS is a disk-based service that provides low-latency storage infrastructure for physics use cases. EOS manages over 120 PB of disk storage at the time of writing this chapter. Access to the open data is provided by the XRootD[13] protocol and is well suited to situations with many concurrent users, a significant fraction of random data access, and a large rate of file opening. The XRootD protocol allows *ad-hoc* streaming of necessary portions of data during an analysis, saving potential users of the large datasets from having to preemptively download the complete datasets before starting an analysis.

In the spirit of openness, development work for the Portal is conducted publicly on the code-sharing platform, GitHub.[14]

4.2. *Cataloguing data and analysis tools*

The CERN Open Data Portal has been developed to serve two main use cases: education and research (see more under Section 5). This is reflected in the information architecture of the Portal itself. It should be noted that special emphasis has been given to providing contextual information that is needed to understand and reuse the data and accompanying software.

On the Portal, each dataset has its own record. Related records are aggregated in dedicated collections, such as those for "Primary datasets" or "Derived datasets". This aggregation, which is standard practice in digital libraries, has been adopted to research data.

[12] https://eos.web.cern.ch/.
[13] http://xrootd.org/.
[14] https://github.com/cernopendata/opendata.cern.ch.

Table 1. The MARC metadata model extended to describe the CMS primary datasets. Not exhaustive, showing some fields only

MARC Tag	Description	Example Value
001	Record ID	1
024	DOI	10.7483/OPENDATA.CMS.A342.9982
110	Corporate author	CMS Collaboration
245	Dataset title	/BTau/Run-2010B-Apr21ReReco-v1/AOD
250	Release	CMSSW_4_2_1_patch1
256	Dataset description	2916 files, 25423849 events, 2678962237287 bytes
556	Validation information	(link to list of validated runs)
567	Selection information	(description of how events were selected)
581	Usage instructions	(link to how to guide)
583	Certification	(description of certification and links to publications)
856	File information	(links to ROOT files stored in EOS)
980	Collection	CMS-Primary-Datasets

Further, metadata — standardised according to MARC[15] — are provided within each record. Adopting the practices developed by partner information platforms in high-energy physics (e.g., INSPIRE and CDS), standard metadata fields were populated. However, various additional metadata fields needed to be created to make sure information about data are standardised and comprehensive across all the datasets available via the Portal (see Table 1).

The information provided was vetted to be comprehensive enough for future reuse of the data. Certain fields, such as the names of the authors, the publication year, title of the dataset, name of the publisher, were made mandatory for the sake of minting DOIs.[16]

[15] https://en.wikipedia.org/wiki/MARC_standards.

[16] See http://doi.org/10.5438/0008.

The metadata are currently stored as XML files, although it is planned to move to JSON schemas with the next release of Invenio (version 3). JSON schemas will provide more flexibility and interoperability with other platforms.

The data and metadata on the Portal are provided under a Creative Commons Public Domain Dedication (CC0). This liberal choice should enable any kind of reuse and exploitation of these datasets. Similarly, the software releases available via the Portal are also published under Open Source licences. The respective licences are displayed on the individual records on the Portal.

Two example records for primary data are shown in Fig. 2 for basic metadata, alongside the Datacite Metadata schema and Fig. 3 for more specific information such as validation procedures or limitations of the content displayed.

4.3. *Preparing user guidance and documentation*

As mentioned earlier, performing a physics analyses with the open data from the LHC is far from trivial. To lower the barrier of entry

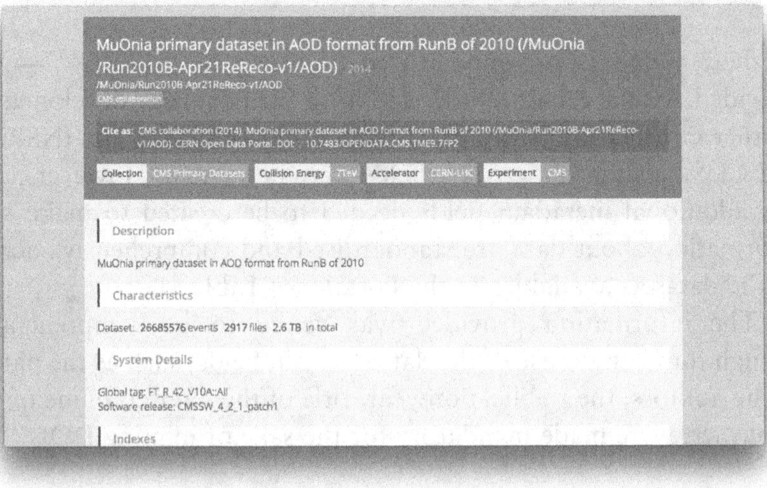

Fig. 2. Primary data record on the CERN Open Data Portal

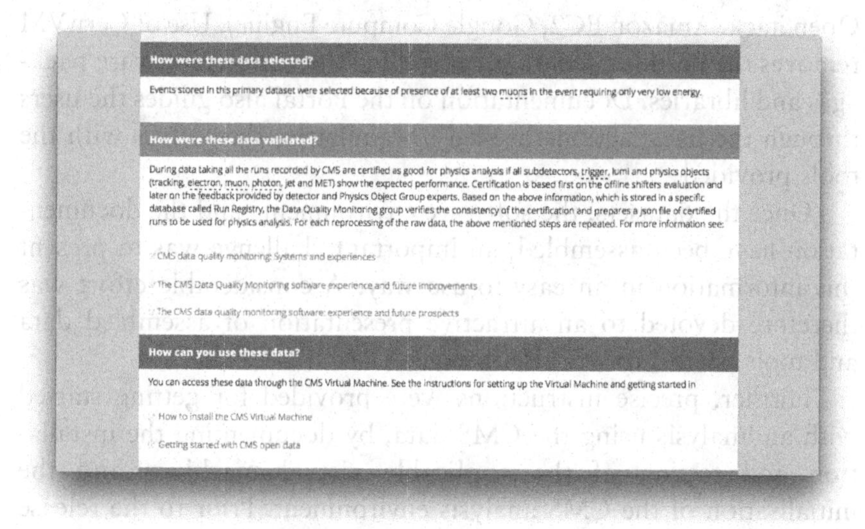

Fig. 3. Detailed metadata is provided to enable users to reuse the data easily

and make the datasets meaningful for students, citizen scientists and members of the wider particle physics community, clear and thorough documentation has been provided, either on the Portal itself or via external resources stored on CERN webpages.

Performing a high-level analysis on the datasets requires the use of a special software framework, known as ROOT,[17] which was developed at CERN. In addition to ROOT, detector-specific software packages are needed to analyze data from the different LHC collaborations. For example, CMS uses a framework called CMSSW.[18] To streamline access to these tools, the software environments necessary to perform a physics analysis on the LHC datasets are captured using the CernVM[19] technology. The CernVM Virtual Software Appliance provides a complete, portable and easy-to-configure user environment for developing and running LHC data analysis locally or on institutional and commercial clouds (such as

[17] https://root.cern.ch/about-root.
[18] https://cms-sw.github.io/.
[19] http://cernvm.cern.ch/.

OpenStack, Amazon EC2, Google Compute Engine). Use of CernVM removes the need for separately installing the various software packages and libraries. Documentation on the Portal also guides the users through the basic actions needed to familiarise themselves with the tools provided.

Once the datasets, the software environments and the documentation have been assembled, an important challenge was to present this information in an easy-to-use way. A considerable effort was therefore devoted to an attractive presentation of assembled data and tools needed to work with them.

Further, precise instructions were provided for getting started with an analysis using the CMS data, by documenting the installation and testing of the applicable virtual machines and the initialisation of the CMS analysis environment. Prior to the release of the data, these steps were extensively tested on a variety of hardware and software configurations. A validation report, based on this testing, was also produced and made available on the Portal.

Yet another barrier, one that might not seem obvious to those used to working with particle physics datasets, is one of vocabulary. This is not limited just to the terminologies within particle physics; several terms are used in CMS-specific contexts, referring to the analysis software or to the information recorded in the datasets themselves. A Portal-wide glossary[20] was therefore created; terms included in the glossary are highlighted on all pages on the Portal, and hovering on the term displays the definition. This helps provide contextual information within the documentation.

5. Use Cases

The division of the CERN Open Data Portal into two top-level sections — Education and Research — reflects the two main anticipated use cases. The division also reflects the respective levels of physics and software knowledge needed to use the content in each

[20] http://opendata.cern.ch/glossary.

section. Primary datasets from the LHC detectors found in Section 5.2 reflect the complexity of the detectors themselves. To be able to extract something meaningful from these datasets one needs at least: knowledge of the physics involved, familiarity with the data format, knowledge of the detector and its performance, a software environment and API specific to the experiment, and software expertise. These are significant hurdles for the general public. The material found in Section 5.1, on the other hand, has already found use by the general public in many cases and is more easily accessible.

5.1. *Education*

The first and most enduring use of public data from the LHC is in education, particularly in programmes designed to teach particle physics to students of high school age (in so-called Physics Masterclasses described in further detail below). Before open data policies were in place, requests for data to use in these programmes were done on an ad-hoc basis. The success of the educational programmes was one of the favorable factors considered by the LHC collaborations when further data releases and open-access policies were considered.

The main content under the Portal's Education section consists of derived datasets (mentioned in Section 3.1) and two browser-based applications for exploration of the derived datasets. Derived datasets are reductions of the primary datasets in the sense that only part of the information in the primary datasets is kept and/or only some of the collision events in the primary datasets are selected. Very often the derived datasets are produced in open, human-readable formats such as CSV, XML and JSON; this is possible when only some of the information from the primary datasets is kept. Software for production of derived datasets is included in the Portal, as are code examples for visualisation and analysis of the datasets.

The actual content and level of complexity of a particular derived dataset often depends on the intended audience. Open data as distributed in derived datasets find widest use in the aforementioned Masterclasses. Here, the intended audience are high school students.

The purpose of the masterclass is to teach students a bit about particle physics and the detectors used to study it, as well as to give them a sense of what it is like to analyse data and to obtain a result. In 2016, the International Particle Physics Outreach Group (IPPOG) Masterclasses coordinated by QuarkNet and TU Dresden were held in 46 countries with over 10,000 students participating between February and April.

A dataset used in a masterclass is prepared by a physicist with a specific analysis in mind. For example, both ATLAS and CMS have exercises where the students are meant to estimate the ratio of positively charge *W* bosons to that of negatively charged *W* bosons. Therefore, a sample of *W* bosons must be extracted from the primary datasets. These datasets are distributed as either CSV or XML files. A typical masterclass day starts with a short lecture on particle physics, how the detector works and how to conduct the analysis to follow. The students view up to 100 events in an interactive collision visualiser (called an event display): an application that visually renders the collisions and the detector. With the display, the students attempt to determine whether or not the event displayed is a W+ or a W−.

Direct exploration of the data is also possible on the Portal itself, via an interactive browser-based event display and histogram tool. The event display visualises derived datasets from CMS; a screenshot can be seen in Fig. 4. The histogram application allows the user to examine the distributions of variables found in several of the derived datasets. An example is shown in Fig. 5.

5.2. *Research*

Open access to the research data will, in the long term, allow the maximum exploitation of their scientific potential. The potential users of the Research collection are experimental and theoretical physicists who were not members of the collaboration collecting the data as well as collaboration members who might want to return to the data long after they were recorded. Also, these data can be of use

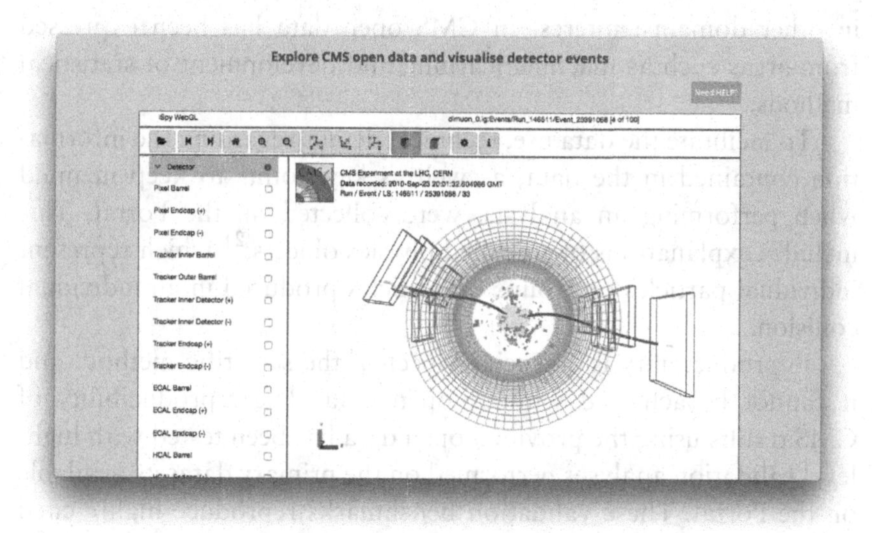

Fig. 4. The browser-based event display found in the CERN Open Data Portal. An event derived from the DoubleMu primary dataset is shown

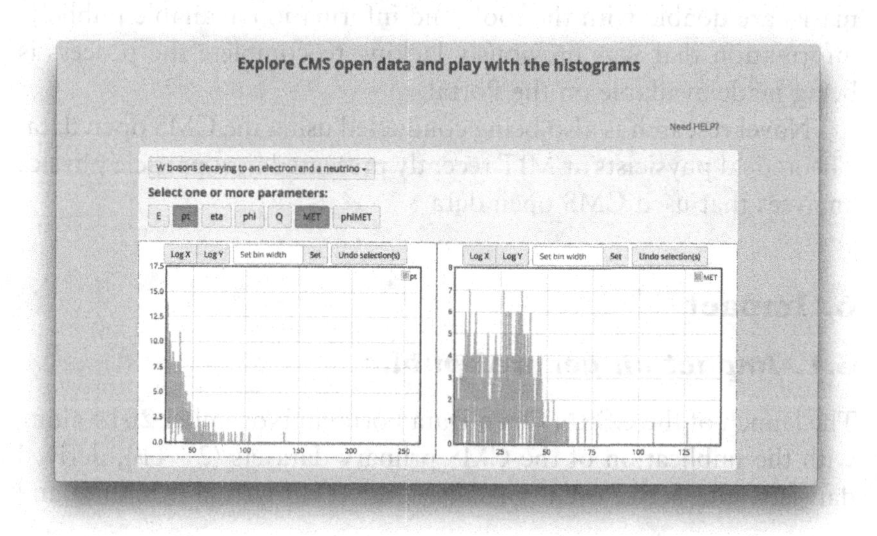

Fig. 5. A screenshot of two histograms created using the application found in the CERN Open Data Portal

in other domains: interest in CMS open data has been expressed from areas such as machine learning and development of statistical methods.

To facilitate the data use, technical details regarding the information contained in the data, as well as key points to keep in mind when performing an analysis, were collected on the Portal. This includes explanations of so-called physics objects,[21] which represent individual particles or groups of particles produced in an individual collision.

Reproducibility is a crucial aspect of the scientific method, and it cannot be achieved without open data. The reproducibility of CMS results using the provided open data has been tested with high-level validation analyses performed on the primary datasets available on the Portal. These validation benchmarks reproduce highly cited analyses that CMS published with a similar data volume.[22] Although these analyses are not too complicated, they nevertheless have interesting physics contents, and offer the possibility of comparison with data at other beam energies in the future. All steps in these benchmarks are doable with the tools and information available publicly; information that was previously lacking to complete the process is being made available on the Portal.

Novel research is also being conducted using the CMS open data. Theoretical physicists at MIT recently presented new particle physics analyses that used CMS open data.

6. Impact

6.1. *Impact on general public*

The launch of the CERN Open Data Portal in November 2014 along with the publication of the CMS primary datasets (27 TB), derived datasets, interactive event visualiser, the CMS virtual machines and

[21] http://opendata.cern.ch/about/CMS-Physics-Objects.
[22] See the validation record http://opendata.cern.ch/record/460 and the corresponding CMS publication http://dx.doi.org/10.1088/1748-0221/7/10/P10002.

the analysis software examples was met with considerable interest from the general public.

The launch was covered by major news networks and on social media, with many articles inviting readers to run an LHC physics analysis "in your living room". In the month following the release, more than 80,000 people visited the Portal, which is several times larger than the active particle physics community (estimated at around 20,000). Most of the visitors spent time browsing around the site and using the provided interactive tools. Several thousands of visitors explored the provided data by downloading the virtual machines and software to study the datasets.

Two weeks after the launch of the Portal, an "Ask Me Anything" session was organised on Reddit,[23] where readers could ask questions about open data and open science practices at CERN. The Reddit AMA session attracted considerable attention and led to another peak in visitors to the CERN Open Data Portal.

The public interest in the released open data generated some non-physics use case scenarios, such as using the large datasets as a robust input for digital forensics studies related to security in cloud computing. It is enlightening to see possible applications outside of the primary focus of the data, which is physics analysis.

6.2. *Impact on scientists*

When the CMS Collaboration first drafted its open data policy in 2012, the concept of open research data was entirely new in high-energy physics. The researchers — correctly — see the published result as the goal of the research work, and publishing data on their own was seen by some as unnecessary, premature or meaningless. Several concerns were expressed within the collaboration, mainly addressing the potential additional workload for collaboration members in case someone presents contradictory or false results based on the open data from CMS. It was, however, considered that the benefits of open data, both in terms of preserving their

[23] https://redd.it/2nxwkb.

scientific potential and in enabling their use in education, overcome any potential risks.

The high-level open data from CMS — in the "primary datasets" — are in the same format as the ones used within the collaboration for physics analysis. There is no simplification or curation of these data for public release. However, in order to prepare these datasets in the first place, the raw data from the detectors must be "reprocessed" and the same version of the software used in the reprocessing must then be used to perform an analysis on the datasets. The only viable solution for releasing the reprocessed primary datasets with the appropriate software versions was to make them available as part of the CMS Collaboration's legacy data. The legacy data for the 2010 data collection period are reprocessed in their entirety with a single version of the analysis software. In this way, there is no need for additional reprocessing and validation just for a public release, a time- and resource-consuming process. The exact provenance details (such as configuration parameters) can be directly extracted from the CMS internal data description systems.

With that being said, finding the human resources to provide adequate instructions and documentation is quite demanding for a collaboration during active data taking, with all the resources focused on collecting and analysing new data. The experience of the first data release from CMS was instructive: even though preparations for the data to be released were started with a fairly short delay after the final reprocessing of the legacy data, and considering that some of the last publications with these data were done just a bit earlier, finding the correct set of instructions and documentation was not easy for many simple reasons. For one, the people in charge had moved to different positions within the collaboration. For another, the documentation had been updated and, although much of it was version-controlled, finding the correct set of instructions corresponding to the exact software release is not as obvious as it may seem. Having learnt from this, CMS is preparing for smoother data releases in the future by recording all the necessary instructions for data analysis with the data that are in active use.

Beginning preparations for releasing open data well in advance has had a major impact on the long-term data- and knowledge-preservation efforts within the CMS Collaboration. Curating the instructions on how to use the data puts the emphasis on the most difficult area in data preservation: the preservation of knowledge, which at the time of the active use of the data is such an integral part of everyday work that it may not even have been explicitly noted down.

The overwhelmingly positive feedback received following the first data release in 2014 has convinced the CMS Collaboration of undertaking further regular data releases.

Index

Printed in the United States
By Bookmasters